BRITISH RAILWAY
GOODS WAGONS IN COLOUR
For the Modeller and Historian

Robert Hendry

Midland Publishing
Limited

British Railway Goods Wagons in Colour
Robert Hendry © 1999

ISBN 1 85780 094 X

First published in 1999 by
Midland Publishing
24 The Hollow, Earl Shilton
Leicester, LE9 7NA, England.
Tel: 01455 847 815 Fax: 01455 841 805
E-mail: midlandbooks@compuserve.com

Design concept and layout
© Midland Publishing.

Midland Publishing is a division of
Ian Allan Publishing Ltd.

Printed in England by Ian Allan Printing Ltd
Riverdene Business Park, Molesey Road
Hersham, Surrey, KT12 4RG

CONTENTS

Preface & Acknowledgements	3
The Wooden Goods Wagon	4
Steel Takes Over	9
Freight Depots	14
An Independent Freight Line	19
Steam Age Freight	21
Transition Era Freight	27
When Coal was King	30
Iron Ore, Stone, Sand and Cement	36
Open Merchandise Wagons	42
Vans	46
Cattle Wagons	53
Tank Wagons	54
Bulk Commodity Wagons	58
Lows and Flats	60
Rail and Timber Wagons	63
Specially Constructed Vehicles	66
Departmental Stock	69
Cranes	76
Brake Vans	77
Appendices: Headcodes & Block Codes etc	80
Selected Freight Stock Drawings	85
Glossary	96

Above: The title page picture is of a steam age freight train. In this scene, taken at Wellington in Shropshire on 30th June 1976, almost eight years after the end of standard gauge steam on BR, the motive power is a diesel, 08 590, but the train, a trip working from the Army depot at Donnington on the stump of the closed LNWR line from Wellington to Stafford, recalls steam age operating practices with its mix of wagons. These include a Tube, several steel Highs, some Conflat As and two Warwells. The traditional steam age goods train and its wagons, survived well into the diesel era.

Front cover captions:

Bottom left: For the first 100 years of the railway era, all-timber wagons prevailed. A few lasted into the 1960s. S56046, a London & South Western Railway brake van, seen at Ryde in the Isle of Wight in October 1966, had outlived most of its contemporaries.

Bottom right: The 16 ton steel mineral wagon was the most numerous type on BR, most being to Diagram 1/108, of which about 200,000 were built. Wagons had a rough life, and on B238031, seen at Moira in the East Midlands coalfield on 26th August 1973, the headstock is tied on with a couple of lengths of string!

Main picture: BR Standard class 5 4-6-0 No 73016 passes beneath one of the LSWR low pressure pneumatic signal gantries at Basingstoke on 29th October 1965. Until the 1960s, a wide variety of motive power could be found on freight trains, ranging from 0-6-0s dating from Victorian times, through dedicated freight engines such as the 8F, to mixed traffic engines, such as the Standard class 5. In late steam days, Britannias and even first line Pacifics, such as Duchesses and A4s were used. The freight train offered almost unlimited variety.

Title page: It is 30th October 1965, and steam on the Western Region will be gone in a matter of weeks. Stanier 'Black Five', No 45288, heads south through Banbury with a class 6 freight, its precise nature being revealed by the headlamps. The class 6 was a limited load train, or had at least four vacuum braked vehicles at the front, to provide added braking power and permit higher running speeds, for the problem with the traditional freight train was not starting or accelerating, but stopping. Even four wagons added markedly to braking powers. No 45288 had been a Willesden engine, later moving to Rugby. By the early 1960s, livestock usually moved by road and this was the last cattle wagon we saw conveying livestock. Cattle wagons were marshalled at the head of a train to minimise the snatch when a loose coupled freight train starts, for with three link couplings, the engine can be moving at quite a speed by the time the slack is taken out and the rear wagons snatched into motion. Whilst goods guards had to accept jolts on starting, they were only guards, so ranked low in the order of creation. Cattle ranked higher, as the farmer might demand compensation if they were injured. Exceptionally, the cattle wagon is conveying horses, which normally moved by horse box on a passenger train. These are either underprivileged horses or are making their last journey. The BR Diagram 352 (12 ton) and Diagram 353 (8 ton) wagons, despite the notional difference in carrying capacity, were virtually indistinguishable from one another, sharing the same overall length, width and wheelbase, the only significant difference being a 3 inch greater height for the larger wagon. The second wagon is a BR plywood bodied van with hinged doors and corrugated ends, a common design, whilst the third vehicle is an SR designed van, distinguishable by the elliptical roof. The difference in the shade of bauxite on the wagons merits comment. To the right is the girder framework of one of the gas holders at Banbury gasworks, which was conveniently sandwiched between the Great Western and LNWR stations, and was served from either.

PREFACE & ACKNOWLEDGEMENTS

Another book on Goods Wagons! In the 1960s, information on freight stock was confined to the odd article in the model railway press. The first serious work for the enthusiast was *British Goods Wagons from 1887 to the Present Day* by Essery, Rowland and Steel. Jim Russell followed with several GWR wagon books, and LMS, LNER, Southern, BR, some pre-grouping, private owner and tank wagons have been covered. Most have been splendid, and as a modeller, I have welcomed their efforts, as it has allowed me to transform our freight stock. Despite a shelf of excellent wagon books, the number of colour views was minimal, and I realised that the most comprehensive colour coverage of wagons was my own *Rails in the Isle of Wight - A Colour Celebration*, which contained 20 views of IoW freight and departmental stock. I came to realise that I used the wagon books for construction, but for in service details, I used the colour archive taken by my father and myself. A discussion about a proposed freight title with Tom Ferris and Chris Salter of Midland Publishing and graphic designer Stephen Thompson led to this book.

Stephen, as a modeller, wanted a colour record of stock for livery and weathering. Chris wanted a description of the different types of train, known to signalmen by their bellcodes, 3-2; 4-1; 3-1-1 and the rest. It is over 30 years since steam hauled freight operated on British Railways and whilst the traditional hand braked or vacuum braked wagon outlasted the steam loco, they are now supplanted by modern high capacity wagons, block trains and dedicated terminals. As with the vehicles, I frequently used our colour archives for such details. Out of the ten train classes in BR steam days, A-K (excluding I), later classes 1-9 and 0, two were passenger, and the rest were parcels, freight or light engines.

Whilst an unfitted coal wagon with a 9 foot wheelbase was to be found on a coal train, pick up goods, or slow freight, such stock never worked on fast, fully fitted freights. Many modellers do not realise this, so although their locomotive may be perfect, their train may be absurd. It is helpful to study freight depots, which have been ignored in previous colour books, where engines have pride of place.

We agreed what we wanted, and I had to create it. That was when the problems began ! In our archive, I had over 500 suitable views from the 1950s onwards, but in covering so broad a subject, a personal collection is seldom enough, but when I asked around, one leading manufacturer replied, 'We are very sorry we cannot supply any colour photographs for your book; however if you ever need any black and white photographs, we will be happy to oblige'. Until the 1970s, black and white prevailed for commercial or enthusiast purposes. My father, in taking colour of freight stock, was in a minority within a minority. Fortunately, a few people have turned up gems, but the heart of this book is the colour record started by my late father, without which it would not be possible.

The stock covered ranges from BR standard designs, through that built by the Big Four, to pre-grouping types, for unlike locomotives which stuck to regional or company boundaries, wagons were more prone to wander. A GWR devotee could build a good layout without knowing anything about LMS locomotives or carriages, but long before nationalisation, the pooling of wagons meant that the basic wagons or common user stock, was randomly distributed. In BR days, even the limited 'non common user' restrictions ended. Private owner stock was widely used for the carriage of coal, chemicals, petrol and minerals and examples are included. Private owner vans were rare, but again we were fortunate. The Isle of Wight provides several wagons, as elderly stock survived there long after it was extinct elsewhere on BR. Another area where archaic stock persisted was Ulster. Although one of the principal railways in Northern Ireland, the Northern Counties Committee, was a part of the LMS (inherited from the Midland Railway), many authors ignore it. As the NCC was a part of the British Transport Commission from January to March 1948, and thus a part of the Railway Executive, I have included some LMS NCC wagons to

redress that balance and because many displayed design features which had vanished in England in Victorian times of which I know of no other examples recorded in colour. Perhaps the most unlikely wagon we have illustrated was built by the LNER in the 1940s using a Midland Railway design of 1919, under an LMS Lot number, for use on the NCC.

The earliest colour photographs date from the 1930s and show how shabby freight stock was at that time. I wish there was a wider selection, as wagons from the 1870s and 1880s still existed, along with half a million private owner coal wagons, but as Ron White of Colour-Rail says, 'if the number of pre-war colour views of Duchesses can be counted on the fingers of one hand, what hope is there for wagons!' Even so, Ron provided some gems. Chris Nettleship and Mike Hodgson have also filled some of the gaps. Most wagons are in service, but a few preserved vehicles appear. I would have preferred a pre-war colour photo of a GWR Mica, but without a time machine, we are limited to what exists! Thanks go to Clive Partridge, whose 30 years of railway service enabled him to explain many anomalies, for example why a train of loaded hopper wagons had four empty 16 ton steel minerals at the head, or why the penultimate vehicle on a sand train was an ordinary goods van.

This book is dedicated to my wife, Elena Aleksandrovna Plotkina. When I began work on it, we had not met. Later, it became a feature of our romance, Elena's initial disbelief that anyone would want to write a book about goods wagons, being overtaken by the realisation that whilst visiting the Crimea to meet her family, I had taken pictures of former Soviet trolley buses and railway wagons. Surprise eventually changed to encouragement, 'Robert, you had better work on your book'.

Below: The standard 5in x 4in wagon label was customarily printed on buff (or white) Manilla 1½ sheet card, but thicker or thinner examples are to be found. Origin and destination might require to be filled in, or one or both could be pre-printed in the case of regular traffic flows.

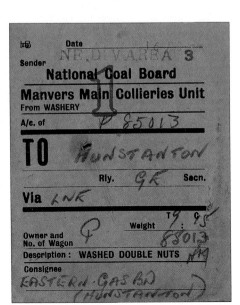

THE WOODEN GOODS WAGON

Top right: The earliest railway wagons owed much to the traditional skills of the village cartwright, who had built vehicles for farmers and coal owners. Instead of employing wooden wheels running on rough cart tracks, they used iron wheels running on fish bellied rails, but as with the farm cart, the body was low mounted without springing, and sat low between the wheels. The chauldron wagon which was familiar to George Stephenson as a young man remained in use in the coal fields of the north east until the decimation of the coal industry in the 1970s and 1980s. Mike Hodgson recorded these examples at Beamish on 22nd November 1974.

Right: One of the earliest railways in the country, the horse-drawn 16 mile Stratford & Moreton Tramway, was authorised in 1821 and opened on 5th September 1826. It was one of the earliest lines built to 4ft 8½in gauge. The line was absorbed into the Oxford, Worcester & Wolverhampton Railway, and then into the GWR. The northern end became superfluous following the opening of the OW&W line in 1859. It remained in use as a horse-worked byway until about 1900 and the tracks were not lifted until the First World War. One of the Stratford & Moreton Tramway wagons was preserved in the open at Stratford. The cast iron wheels were produced by Smith & Willey, Windsor Foundry, Liverpool, and are an early example of a split spoke cast iron wheel. The slight curve in the spoke was to reduce stress problems during cooling. The lack of springing was acceptable whilst vehicles moved at walking pace, and the wagon is as simple as possible,.This and similar vehicles on colliery tramways throughout Britain were the starting point for the railway wagon. The livery, which is a mixture of brown, probably creosote, and sun-bleached wood, is not what we associate with preservation and the manner in which the nearest dumb buffer has rotted away completely, suggest the wagon was not well looked after at this time, but indicates how a wooden wagon weathered in service.

Above: Wagons endured a hard life and BR embarked on a massive renewal programme, which was to see *half a million* new wagons in service by 1960. Older stock rapidly vanished, but it was still possible to see elderly equipment throughout the 1960s on the National Coal Board lines. The most remarkable survivors were in the north east, where slope sided timber hopper wagons had evolved in Victorian days for use on the private coal railways and also on the North Eastern Railway itself. After an earlier phase with 10/12 ton wagons, the NER opted for a 20 ton design from 1903, and by the grouping had over 17,000, for alone amongst the mainline railways, the NER actively sought to carry coal in its own wagons rather than private owner wagons which were the rule elsewhere. The LNER added a further 7,000 and many entered colliery use after their main line days were over. Others were built new for the coal companies. SRW10715 and SRW10540 at NCB Lynemouth on 10th June 1967, carry the prominent wooden end stanchions of the earlier 20 ton hoppers, which originally extended below the body to form dumb buffers so they could work with the archaic chauldron wagons. Main line examples built after 1923 had steel T section stanchions, and from the 1920s the LNER cut off the wooden extensions on the former NER stock. The nearest locomotive, No 49, a Bagnall, work's number 2750, was built for the War Department in 1944. No 25, the other locomotive in the picture, was Hunslet, No 3217 of 1945, also a WD engine. *Colour-Rail*

Right: The NCB was concerned with the movement of coal, so for a balanced wagon fleet dating from Victorian days, which displayed design features which had been considered archaic even in the 1870s, it was necessary to visit Northern Ireland, where, from 1948, the Ulster Transport Authority operated the former Belfast & County Down Railway, the Great Northern Railway (Ireland), and the Belfast & Northern Counties Railway. The latter had been taken over by the Midland Railway as its Northern Counties Committee in 1903, by the LMS in 1923, and after Nationalisation, was a part of the British Transport Commission from January to April 1948. NCC wagons were briefly a part of 'The Railway Executive' so come into our story. This astonishing survivor, UTA 2569, is a BNCR designed 10 ton open, and was in use at Adelaide shed in Belfast, on 25th August 1966. Compared to the Stratford & Moreton or chauldron wagons, the body is carried above the wheels, permitting a higher carrying capacity, but the sides still rest directly on the solebars, rather than on a separate curb rail, to permit a wider body. The axleguards, or W irons, as they were once called, are bolted to the outer faces of the solebars, a practice archaic in England by the 1850s, yet standard BNCR policy to the end of the company's existence! Once mounting the body above the wheels evolved, the convenience of side doors became apparent, and builders provided structural strength by bolting the side and end planks together on open wagons with right angled straps. Drawings of the 1840s show this, but by the 1860s, right angle corner plates had prevailed. A few BNCR three-plank opens were rebuilt with corner plates, but many, including No 2569, retained this early feature until they were broken up in the 1960s! By the 1850s, iron side stanchions were provided on each side of the drop door. The BNCR used wooden posts bolted to the planks and secured to the solebar by iron straps. Crude brakes evolved with an iron plate bolted to the solebar carrying the pivot point of a brake lever which worked via a right angle crank to a brake shoe. It was as simple as possible, but provided only one braked wheel per wagon. By the 1860s, though wagon builders offered a superior system with a central V hanger, which we will illustrate later, the BNCR remained faithful to the archaic single-shoe brake. This wagon carries the grey livery used by both the LMS NCC and the UTA for its unfitted stock.

Centre right: Another area where ancient wagons lingered was in the Isle of Wight. No 425S was built for the Isle of Wight Railway in 1865 by the Kirkstall Forge Co of Leeds and is in Ryde St John's works yard in April 1962. The crane could lift 2 tons and when in use, a counter weight box was wheeled out along the tail to the appropriate distance. Rather than a conventional hook and three link coupling, it has a shackle and 5 short links, an arrangement common in the 1860s but archaic by 1900! The vehicle to the left, S28341, is a Diagram 1369 London Brighton & South Coast Railway 5-plank drop door open. It has just received a major repair, which had left only a few of its older planks. Unlike the BNCR open, it has conventional corner plates and iron strapping. The body is supported away from the solebars by means of a curb rail. By the 1890s, 5 plank general merchandise opens were being produced to this style by most companies. The last completely new all timber bodied 5-plank opens to be built by BR appeared as late as 1952. It was only in details that they are different from this LBSCR design. The wagon has the V hanger brake system, with brake shoes worked by push rods on both wheels. A brake lever guide can be seen on the far side of the wagon just below the left hand end stanchion, so it is fitted with brakes on both sides.

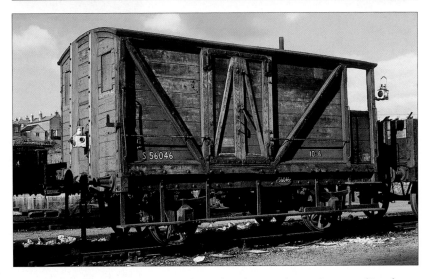

Above: Although the majority of wagon builders adopted corner plates and iron stanchions for open wagons, timber framework remained the rule for covered stock. Brake vans in particular tended to be externally framed. S56046 was a London & South Western Railway Diagram 1541 Standard Road Van. Most of its contemporaries vanished in Southern Railway days, but this was one of a handful of vehicles sent to the Isle of Wight, where it is seen at Ryde on 29th October 1966. It was one of three antiquated designs which survived into the 1960s. The Road Van concept meant that as well as providing brake power, it conveyed less-than-wagon-load traffic to wayside stations, hence the double doors for unloading purposes. The wagon is in the brown livery with red ends adopted by the Southern Railway for brake vans and retained by BR on the IOW section.

Right: As we have seen, stock was sometimes sold to private owners, permitting a few elderly vehicles to survive into modern times. This 4-wheel brake, which came from Cannock Wood Colliery, is at Blists Hill in 1976, and typifies nineteenth century construction techniques. It has split spoke wheels, oil axleboxes and is similar to the LSWR road van illustrated on the previous page. At this time, most companies built single ended brakes with a veranda at one end only, the modern symmetrical brake was a twentieth century development.

Centre right: By the 1880s, the Belfast & Northern Counties Railway standardised on an externally framed 10 ton van with cupboard doors. As with the BNCR opens, the sides rested directly on the solebars, giving an antiquated appearance, as did the external W irons. The original short leaf springs have been supplemented by longer springs secured to separate anchor points. This was to improve the riding and enable the vehicle to operate as 'tail traffic' on passenger services. Unfitted vans do not seem to have received this improvement. Unlike the wagons on either side of it, which retained the anchor plate for the brake lever and a single wheel brake, UTA No 1612, seen at Adelaide Yard on 21st August 1965, has the more modern V hanger and Morton brakes. The Morton system permitted brakes on one side of the wagon to be operated by levers on either side of the vehicle, saving the need for a full set of brakes on both sides. As the brake is worked by depressing the lever, so rotating the brake shaft, one lever must be fitted with a dog clutch, which comprises a pair of toothed wheels. The wagon is in the brown livery adopted by the UTA for fitted stock.

Bottom right: Traditional though the previous wagons have been, this van is an even more striking example of conservatism. It looks like a typical Midland 12 ton van, and is to a 1919 MR design in which the diagonal bracing, instead of being in the familiar V form each side of the door, comprises two parallel diagonals, sloping away from the headstocks. Only 200 were built by the Midland, but over 20 years later, the design was revived when the LNER built 100, to LMS Diagram D2074, for the LMS NCC where they were numbered 2401 to 2500. The design was so obscure that Bob Essery, in his magnificent two volume history of LMS wagons, wrote, 'Regrettably, no pictures are known to exist'. The construction of a pure MR design by the LNER 20 years after the Midland ceased to exist is remarkable. Unlike normal NCC stock, which carried a rectangular wagon plate, lettered BNCR, MR.NCC, or LMS.NCC, the LNER fitted standard LMS D-shaped wagon plates. Although No 2489, at Adelaide in August 1965, is not fitted, so should be in grey livery, traces of the LMS NCC lettering are visible on the faded bauxite paint, so it must have entered traffic in bauxite. Presumably the LNER painted it bauxite on the grounds that the LMS had changed from grey to bauxite for all stock, regardless of its brake system, in the 1930s. Its last use was as a waste paper van, taking surplus railway paperwork for disposal!

Above: The earliest known colour view of a large freight yard is this 1936 study of Rugby up yard from the wooden bridge. We are looking towards Rugby No 7 box and the divergence of the Leamington, Birmingham, and Trent Valley routes. A domeless Stanier 5XP 'Jubilee', No 5564 *New South Wales* heads an up express towards Euston. An engine is blowing off in the down sidings on the left, and there is a second engine on the open wagons in the dead end siding. Hump shunting is in progress in the up sidings on the right.

The brake van, to the right of the 'Jubilee' is a 20 ton Diagram D1940 van built in 1933/34, which retained the 20 foot body length of earlier LMS vans, but went from a 12 to a 14 foot wheelbase. It carries the pre-1936 livery of light grey with the letters LMS written large. The numbers are in white on a black panel edged in white with reversed corners.

The van's handrails are white. The three single bolster wagons ahead of the brake van show the use of a match truck for long loads, as the load does not rest on the bolster of the nearest wagon. The furthest wagon is a GC steel framed single bolster, whilst the nearer pair are NER timber framed wagons. The fourth wagon is a sheeted open, though the sheet has subsided into the wagon, and would merely trap rain. The deplorable state of the van in the foreground merits comment, for its 'livery' is grime. A rake of coke wagons occupies the next siding, including two J C Abbott private owner wagons. This was a large Birmingham concern which operated over 800 wagons and was in existence from the dawn of the railway age. Coal was supplied to gas works and other industrial and private users throughout the midlands. The company bought coke in large quantities, mostly for iron manufacturing,

chain making and other heavy industries. The wagons were black with white lettering. The rectangular yellow 'commuted charge' plate is just below the A of Abbott on the nearest wagon. Sandwiched between the Abbott wagons is a Bedwas open, some of which were lettered Bedwas Coal, others Bedwas Coke. The livery is grey with black ironwork, but is so grimy that it looks similar to the black wagons on each side of it. There is a further Bedwas wagon to the right of the 'Jubilee'. Bedwas was on the GWR in South Wales and as J C Abbott & Co are known to have used Bedwas as a supplier of coke, it is likely that this rake of Abbott and Bedwas wagons has come in from South Wales, bringing coke to industrial users in the area. Had the destination been London, it would have been routed via the GWR main line.
H J Stretton Ward

Left: This scene from Colour-Rail of a former LNWR 'Chopper' tank No 6428 at Steeple Grange on the Cromford & High Peak line in September 1943 had to be included for several reasons. Firstly the C&HP, with its inclines and archaic motive power was an incredible freight only appendage of the rail network. Second, the chance to include a chopper tank on a freight in colour was not to be missed, and most important was the leading wagon. My father developed a comprehensive 0 Gauge model railway, and one of the POW wagons which fascinated me as a child was a private owner open from Smeaton, which is south of Edinburgh on the former North British Railway. The word 'Moore', on the side of the wagon stood out, because of the elongated middle letter. As I looked at this slide, there was that elongated middle letter! *Colour-Rail LM7*

Left: By the 1890s, a considerable degree of standardisation had developed through the Railway Clearing House which specified standards for most wagon components. New privately owned wagons were inspected for compliance with the regulations, prior to acceptance. Except on the North Eastern Railway, and to a lesser extent the Midland, railway owned mineral wagons were rare prior to 1948. The few which did exist met RCH requirements, which facilitated sale to private owners later on. This Rugby Portland Cement Company open, displaying faded RPC light grey livery with white lettering shaded black, began life as a MR Diagram 607 coal wagon, of which 5,650 were built from 1911 to 1921. Traces of the MR lettering remain, but the vehicle was so weathered, when photographed at Southam Cement Works in June 1966, that no RPC or Midland number was visible. Intended for loco coal, they were used for many commodities and were fitted with side and bottom doors though they lacked end doors. In contrast to most mineral wagons, with diagonal straps running upwards from the door opening to the top of the corner plate, the Diagram 607 wagons had an internal vertical strap or knee and an extra external vertical stanchion. This, and the deeper fifth plank, made them readily recognisable. As the majority of private owner wagons, other than specially constructed vehicles, were taken into government control on the outbreak of war and transferred to BR in 1948, this wagon, if sold to the RPC before the war, might have been restricted to internal use even then, or perhaps it had been acquired second hand from BR for internal user duties at Southam after nationalisation.

Below: NCB No 399R, at Moira, in the Leicestershire coalfield, on 26th August 1973, makes an interesting contrast with the MR wagon. The ribbed buffer casings and lack of crown plates on the wooden solebar above the axleguards indicate it is to the Railway Clearing House 1923 pattern. The diagonals are more typical than the Rugby Portland Cement wagon, seen below, but unlike many such wagons, the side stanchions are not curved at the lower extremity to overlap the diagonals. Seven planks was seen as standard, but 8-plank versions, as here, were common. The technical term for the planking is *side sheeting*.

As well as side doors, it is fitted with an end door, though this is at the opposite end of the wagon. End door wagons were common in larger private owner fleets as they were suited to the coal hoists used in the majority of coal ports and where the wagon was tilted. One reason for the retention of end tilting rather than rotary tippling was that the earlier oil axleboxes lost their lubricant if inverted, but not if tilted. In NCB days internal user wagons were frequently distinguished by a large diagonal white cross, but the only indication on No 399R is the small stencilled INTERNAL USER ONLY inscription.

STEEL TAKES OVER

Below left: So far, all the wagons illustrated have been timber framed, for most pre-grouping companies remained faithful to timber. However, the GWR was using iron or steel frames by the 1880s. Except for minor changes, such as to a 10ft wheelbase for better riding at speed and the angled bottom door plank to reduce rot in this area, the designs altered little to the end of company days. Wooden bodied opens appeared in 3, 4, 5 and 5½-plank varieties. Latterly minor changes took place to simplify construction. DW34783 was a late period 5½-plank open without a curb rail, so that the ends of the floor planks project under the side sheeting. The side sheeting is secured by heavy cranked stanchions fastened to the underframe. It is fitted with screw couplings, vacuum hoses and Morton brakes and is on the down goods line at Rugby on 16th March 1971 after its transfer to departmental duties.

Below right: Although the GWR was building iron and later steel framed wagons by the 1880s, and the LMS used steel frames as standard, the LNER remained faithful to traditional wooden frames until the 1930s, whilst the majority of private owners preferred wooden frames. Natural conservatism and corrosion problems with steel wagons militated against a change, but the main reason was the ease with which timber wagons could be repaired. In Victorian days, small wagon repair depots sprang up throughout the country. Unlike the large railway works, they comprised two or three sidings, a joiners shop, a smithy and hand tools. With limited facilities, it was remarkable what a skilled joiner could accomplish on a wooden wagon and many private owners were fearful of increased repair costs if they moved away from familiar and well tried designs. Wagon Repairs Limited had depots throughout the country. They were seldom photographed, but this is their yard at Melton Mowbray on the Syston to Peterborough line of the Midland Railway on 30th July 1971. Wagon wheels are scattered about the yard and a timber stack is in evidence in the distance, as is a hand grind wheel for sharpening tools.

Above: Charles Roberts & Co, one of the most prolific of the private wagon builders, advocated steel stock and in 1935 patented a slope sided steel wagon in which the side, instead of being carried out from the solebar on a side rail, was directly mounted on the solebar. It was a steel version of the NER 20 ton hopper but without the hopper doors. The sloped sides increased carrying capacity, whilst the absence of the curbrail simplified construction and reduced tare weight. The drawback was that this 15 ton wagon was taller than the conventional 12 ton wooden mineral wagon, so could not work under the coal loading screens at many collieries. The design

for a 15 tonner, uprated to 16 tons during the Second World War, came in two types. One had the traditional side, end and bottom doors of the normal private owner coal wagon. The other was for use in rotary tipplers at major plants such as steel works and was devoid of doors. In 1939 Stewarts & Lloyds acquired two prototype tipplers, increasing their fleet to 700 the following year. As specialised wagons, they were not requisitioned when war broke out and remained privately owned until their withdrawal in the early 1980s, by which time S&L had become a part of the British Steel Corporation. Former S&L No 9426 is at Syston North Junction in 1975, loaded with coal for

the BSC plant at Corby. The wagon displays the NON POOL description applied to non requisitioned wagons and the CC, commuted charge, symbol. This stemmed from an agreement reached between private owners and the railway companies before the war whereby the owners made a special payment and were relieved of certain routine charges thereafter. The 'circle and bar' registration plate is fitted to the solebar in the regulation position. All private owner vehicles had to carry such plates which indicated the date they had been registered or accepted and the name of the main line company which had approved them for operation on the rail network.

Left: An Ideal Stocks Committee was set up in February 1948 to consider how the motive power and rolling stock fleet should be developed. Their report on freight stock which appeared in 1950 made many important recommendations. Traditional wooden bodied coal wagons were to be replaced as soon as possible with modern steel stock. The committee considered the 16 ton 4-wheelers built in considerable numbers for the Ministry of Supply, older 20 tonners, and a few bogie 40 tonners. They felt that bogie stock was too large for UK conditions, and proposed the largest practical 4-wheeler, which could carry 24½ tons. It was not a good idea, this view of 24½ tonners in trouble at Abercynon in South Wales in July 1973 is symbolic of the troubled life of these wagons. Too large to clear many colliery loading screens, less than 4,000 were built. *Mike Hodgson*

Above: The Ideal Stocks Committee realised that handling facilities might be a problem, so recommended that the 16 ton steel mineral be produced as a stopgap, using a 16ft 6in channel section chassis. Hand brakes were specified. As the Charles Roberts slope sided wagon was taller than the traditional wooden bodied coal wagon, so did not fit under all colliery loading screens and was a patented design on which royalty payments were due, this design was not pursued. However both the LMS and LNER had developed a more traditional steel mineral wagon which became the basis of the BR 16 tonner. Within a few weeks of the report being published, over 50,000 were on order. Soon the 16 tonner became the most numerous wagon on BR, and sights such as this, at Rugby on 28th June 1971, became commonplace. Of the four wagons pictured, the first, second and fourth can be identified. These are B209118, B240817 and B99534 built to Diagram 1/108. They carry the original style of lettering adopted by BR, with a black patch on the left hand end of the vehicle on which was recorded their capacity and number. A smaller black rectangle at the right hand end gave the tare weight. This typical coal yard scene should be of interest to modellers.

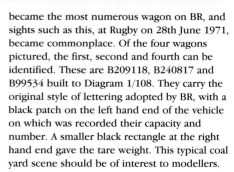

Left: This illustration recalls the problems which could arise in everyday life. The old joke about something being tied on with a piece of string is well known, but when I photographed B238031, a 16 ton steel mineral built to BR Diagram 1/108, at Moira in the east midlands coalfield on 26th August 1973, the headstock was really held on by a couple of lengths of string! The brake lever operates shoes on the other side of the wagon, the Morton clutch in this case being attached to the lever on the same side as the brake gear.

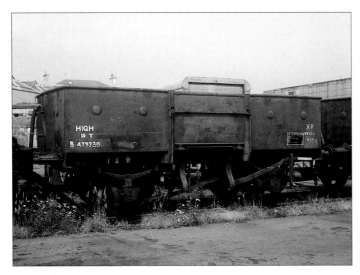

Above left: For open merchandise wagons the Ideal Stocks Committee also temporised, recommending the steel sided LNER 13 ton High, but accepted that for some traffic, where the goods might be stained by contact with rusty body panels, a wooden body was better. In any case, the main workshops were still geared to producing wooden bodies, so it made sense to continue the dual approach. B479730, at Canute Road, Southampton, on 22nd July 1972, is a 13 ton Diagram 1/041 steel bodied High, a development of the LNER design. It was built at Shildon in 1951 to Lot 2195 and is vacuum fitted with the LNER style 8-shoe clasp brakes. In this system the V hangers are offset with one to the right on the vacuum cylinder side and two symmetrically positioned on the opposite side. As built, the steel Highs carried a chalking board between the number data and the door, the four bolts for which are visible on the bodywork. A second panel, also missing, was provided on the end, above the right hand buffer. Unlike the mineral wagons, where a 16ft 6in chassis and 9ft wheelbase was adopted, the standard dimensions for general merchandise wagons were to be 17ft 6in and a 10ft wheelbase. Except for differences in brake gear, with 2 shoe Morton, 4 shoe and 8 shoe clasp brakes of various sorts, the merchandise wagon chassis was highly standardised.

Above right: Although steel was preferred for opens, the Ideal Stocks Committee recommended a 12 ton van with pressed steel corrugated ends and wooden sides on a 17ft 6in vacuum braked chassis with a 10ft wheelbase. There were some interim designs but the archetypal BR van was the Diagram 1/208, typified by B763982, at Banbury in June 1966. It was built at Wolverton to Lot 2414 in 1952 and whilst shortages of materials had led to some vans appearing with plywood doors or sides, B763982 had horizontal body sheeting, and vertical planking on the doors. Although the LMS and LNER standardised on sliding doors before the war, the Ideal Stocks Committee preferred hinged doors. As these could cause damage if they came unfastened in transit, this was surprising. Another retrograde move was that despite vacuum brakes, instanter couplings and two-shoe Morton brakes were specified. This van has been used for something unpleasant, as someone has chalked, 'Dirty Smells', on the door. The fresh paintwork on the door strapping and hinges suggests the ironwork has recently been repaired and given a coat of protective paint. This was usual after such repairs, for if neglected, the bolts would rust, making subsequent repairs harder. Except in incidentals, the BR Diagram 1/208 van of the 1950s does not differ greatly from the 1880s NCC design. This sums up the astonishing stability of railway freight stock design over 80 years.

Right: The main emphasis was on conventional stock and by 1960, over half a million vehicles had been built to BR designs, but with the collapse of less-than-wagon load freight, the loss of perishables, the closure of rural goods depots and decline in the coal trade as householders switched to gas, electricity or oil for heating, BR found its new but traditional wagon fleet far in excess of its needs. In the 1960s rows of surplus wagons accumulated on spare track throughout the system and BR built wagons followed the steam locomotive to the scrapyards in growing numbers. This was Moreton-in-Marsh on 10th September 1971, where the closed Shipston branch had been used for wagon storage. This line had a strange history. It was opened in 1836 as a branch of the horse worked Stratford & Moreton line and was taken over by the Oxford, Worcester & Wolverhampton Railway in 1847. The OW&W fell into the GWR fold, but the Shipston line remained horse worked, as the original Act required. After local pressure, the GWR obtained powers to upgrade it and the line reopened in 1889 as a conventional branch line, closing to passengers in 1929 and to goods in 1960. A short length was retained at Moreton for wagon storage. The nearest wagon, B763919, is a plywood bodied Diagram 1/208 van. This is followed by B850504, a Diagram 1/207 Shocvan built at Ashford in 1950, with the original long thin vertical shock stripes. The third vehicle is a planked Diagram1/208 van which is coupled to a further Shocvan, this time with the later short and fat shock stripes.

Above left: The Ideal Stocks Committee report was conservative, but provided the basis on which BR freight stock evolved over the next fifteen years. However, over that period new types, not envisaged by the committee, were required. One of these was the Presflo 20 ton hopper cement wagon, the prototype being ordered in January 1954. This new design owed little to pre-nationalisation practice. It was gravity loaded, but discharge was pressure assisted by compressed air. B888794, a Diagram 1/272 Presflo built to Lot 3175 by Central Wagon in 1958, is seen at Gloucester in July 1976. 1,891 cement Presflos were built and the design was later adapted to convey salt and power station fly ash.

Above right: From 1948 to 1958, new wagon deliveries to BR ran at not less than 30,000 per annum reaching a peak of 61,479 in 1956. By the early 'sixties, numbers had slumped and were to reach a low of 769

new wagons in 1964! Traditional wagons were no longer wanted, as customers deserted rail for road haulage. New ideas were needed. One was that the next generation of freight stock would be high capacity air braked bogie vehicles, often dedicated to specialised traffics. This scene at Fakenham on 12th July 1971, is a radically different vision of rail freight. Instead of the general purpose railway-owned 4-wheel van, we see a privately-owned wagon dedicated to one commodity. Shellstar 010 was built by the Gloucester Railway Carriage & Wagon Co in 1968 for the start of palletised distribution of fertilisers from the UKF plant at Ince in Cheshire. It was owned by a wagon hire firm, Lloyds & Scottish, whose joint horse and thistle insignia appears on the right hand end of the vehicle. When new they had blue curtain covered sides, quickly replaced by metal cupboard doors, making them similar to a later batch of wagons.

Right: Tar and other liquids were carried by rail from the start in barrels or casks. The earliest rail tank wagon seems to have been built in 1865 and by the 1880s rectangular tanks were used for tar. The first cylindrical tankers appeared in the 1880s and in 1887 the Railway Clearing House issued detailed specifications for their acceptance on the main line railways. The early tank was a slow moving 4-wheeler, the barrel being secured to the chassis by straps, tie wires and end stanchions. In 1944, a new form of anchor mounting was agreed, paving the way for later 4-wheel designs, which reached 35 tons in 1957. In the early 'sixties, a 45 ton gross laden weight was approved, the practical limit for a 4-wheeler. Bogie tankers had been rare, just a handful having been built, but problems with the riding of the 45 ton tankers at the higher operating speeds now desired prompted the construction of a prototype 90 ton glw bogie tanker in 1965.
The 1966 increase to a 25 ton axleload allowed 100 ton glw designs.

Petroleum products were divided into two groups, class A for the highly inflammable liquids such as petrol and class B for less hazardous fuel oils. The livery for class A tanks has been silver or grey for many years, whilst class B tanks were usually painted black to minimise the effect of oil stains. Amoco (UK) Ltd commenced operations from Milford Haven in 1968 with 45 ton 4-wheelers and 100 ton bogie stock. No 85020, seen at Gloucester in July 1976, was built for Lloyds & Scottish by Charles Roberts & Co in 1970, and leased to Amoco. The number is a part of the TOPS computer system. Prior to TOPS, each owner numbered stock in his own series, now each wagon was given a unique reference number as well as a vehicle type code and an owner code in the case of privately owned stock. Amoco tanks were usually coded AMOC, but 85020 was coded LS in recognition of its owners. The contrast between high capacity dedicated user vehicles such as the Shellstar pallet wagon or this Amoco tanker, and the wagons we have seen from the 1950s, is graphic.

Above: Perhaps the most dramatic change in the freight scene was the introduction of the Freightliner concept, which commenced operations in November 1965 between London and Glasgow. Block trains of semi-permanently coupled air-braked bogie flats hurried containers between a limited number of Freightliner termini provided with large unloading gantries. In 1967, there were 46 Freightliner trains daily. By 1977 this had risen to 125 such workings. E3081, a class AL85, later class 85, heads 4D60 north through Rugby on the evening of 11th August 1972. The extent to which Freightliner containers dominated the scene at this time is apparent. The first 30ft box container is in pre-1968 Freightliner livery with a BR double arrow symbol. Under the 1968 Transport Act, the operation was vested in a new company, Freightliners Ltd, under the auspices of the National Freight Corporation. The second wagon conveys three 20 ft containers in the new Freightliners Ltd livery, whilst the leading container on the third flat is in the old livery but minus the double arrows. In 1978 Freightliners Ltd returned to British Rail. Had the railways not invested in modern freight handling methods, the haemorrhage of traffic to the roads would have continued, though the advent of these new freight handling methods spelled the end for the traditional wagon.

Above: The new freight concept evolved rapidly, but the elimination of half a million BR built wagons plus the last of pre-1948 stock, took time. Throughout the 1970s, British Rail operated modern high-speed air braked vehicles, traditional general purpose vacuum braked 4-wheel wagons and hand braked 4-wheelers. On 2nd July 1976, I photographed this scene, as a class 47 backed a long train of loaded coal wagons into the up refuge siding at Hartlebury. Although some of the wagons are vacuum fitted, the presence of grey unfitted wagons near the head of the train, indicates that the vacuum brake, if connected up, was only in use on a small part of the train. A number of accidents on facing points in the early days of railways had imbued operating men, and the Board of Trade, with a horror of facing points, except where unavoidable at junctions or large passenger stations. Connections into wayside goods yards or refuge lines where freight trains could be overtaken by faster passenger services were by trailing connections. A freight train, which was to be overtaken, stopped clear of the points, so that the engine could propel it into the siding. It was time consuming and although this train consists of 16 ton steel mineral wagons and a 1960s diesel, this process would have been familiar to a railwayman of the 1860s.

FREIGHT DEPOTS

Railway wagons existed to move goods from A to B, and unless we know what A and B were like, our understanding of freight operations is incomplete. Goods depot varied from the smallest wayside station with a single siding, to large city freight depots, which could have tracks on two levels, and handled hundreds of wagons a day. Apart from the public goods depots, there were marshalling yards to sort thousands of wagons every day and to dispatch them on the correct train. Large industrial concerns, such as collieries, steel plants, gas works or cement factories had their own private sidings. Wagons were taken into the docks and alongside the quays where ships arrived and departed. Enthusiasts photographed the trains but seldom filmed this aspect of railway operation.

Above right: There could be no better way to start than by looking at a wayside station which had changed little from the 1850s to the 1960s. Llanfair PG in Anglesey is well known for its awesome 57 letter name. At the start of the 1960s the yard retained the early 'star' pattern of sidings radiating from a wagon turntable. This design was used in early yards, when wagons were moved by shunting horse or manually from a turntable to short sidings radiating across the yard. In most places, it had long been replaced by sidings

reached off points, which requires a longer yard, but at Llanfair, limited traffic and an awkward site led to the survival of this relic from the past. In steam days, most wayside stations possessed a hand crane. The 5 ton crane at Llanfair is to the left of the cattle wagons, which are a reminder of how important livestock services once were.

Below: The entrance to the goods yard customarily adjoined the station forecourt, as at Southam Road & Harbury station on the

GWR Birmingham main line, just south of Leamington, seen on 3rd July 1964. The station buildings were matching Brunel Italianate structures with round-headed windows and projecting roofs. By the 1960s, passengers were rare for much of the day, and the driver of the BR dropside cartage lorry, still resplendent in carmine and cream, has parked so close to the front entrance of the building, that it was impossible to use the entrance. This might be seen as inconvenient, but there were no passenger trains for hours, and the

locals would go via the entrance on the platform anyway. The cattle dock and end loading facility are in the centre of the view, and a former GWR goods van, with the Churchward short brake lever to the left. As many later GWR vans had vertically planked doors and conventional brake levers, this may be a survivor from the 1920s, but too little is visible to be certain. The modeller may notice the fire buckets, sagging gutters, with grass growing in them, blue paving bricks on the floor of the cattle dock and the timber buffer stop.

Top right: In 1948, BR inherited 1,886 passenger only stations, 4,815 stations handling passengers and freight and 1,593 goods only stations, or 6,408 freight installations in all. Pre-Beeching and Beeching closures saw the number fall significantly, but as late as 1967, there were still 4,600 freight terminals. By 1977 the number had halved to 2,338. Many passenger lines which lost their services during the Beeching era survived into the 1970s for freight, but then succumbed. They included some delightful rural backwaters in north Norfolk. A long cross country route ran north from Wymondham Junction, on the Cambridge to Norwich line, to the coast at Wells. It was closed completely north of Fakenham in 1964 but remained open as far as Fakenham for freight for a few years. Ryburgh was a tiny station a few miles south of Fakenham, and the yard, with its mixture of grain hoppers and a solitary 16 ton steel mineral, with top flap and main doors open, basks in the summer sun on 11th July 1971. It was a delightful anachronism.

Centre right: Special traffics dominated some areas and nowhere was this more apparent that in the lovely Vale of Evesham, with its multitude of fruit and vegetable growers. Sacks are transferred from C J Howse Ltd's AEC flatbed lorry to a vanfit at Moreton-in-Marsh on 10th September 1971. The number of vans give an idea of the extent of the fruit and vegetable traffic which moved by rail in the early 1970s. A supermarket now occupies the site of the coal yard in the background.

Bottom right: How were wayside stations served? The answer was the pick up goods. This was known to enginemen as a class K (later class 9) and to signalmen as a '3', on account of its bellcode. The pick up trundled from station to station, spending five minutes here, 20 minutes there and an hour somewhere else. It brought 16 tonners to the coal merchants, agricultural feedstuffs for the farmer and sausages and potatoes for the village store. Once the wagon was loaded at a wayside station, it began its journey on the pick up, which would take it to a larger yard, where through freights were marshalled to major destinations. There it would be added to another pick up to complete its journey. BR Standard class 4 4-6-0 No 75067 shunts a pick-up at Brockenhurst in 1964. Brockenhurst is a pleasant town in the New Forest between Southampton and Bournemouth on the LSWR line from Waterloo. The Southern did not use conventional headlamp codes to distinguish the type of train, but a route code instead. There were six lamp or disc positions, one at the base of the chimney, a pair on the side of the smokebox and three above the buffer beam. The code carried by this locomotive, bottom left and mid right, indicates the Waterloo/Nine Elms route to Bournemouth via Brockenhurst and Sway. The leading wagon is a Conflat A with an ice-blue 'highly insulated' type AF container.

Above: Large deliveries of air braked freight stock only began as the steam locomotive was approaching extinction, so the end of steam did not herald an immediate end to traditional freight workings or the demise of pick up goods. Although displaying class A Express Passenger discs, 40 116 has arrived at Skipton, left several wagons and a brake van in the platform, then dropped off four coal wagons in an adjoining siding, the nearest two being vacuum fitted TOPS coded MCV 16 tonners without drop flap doors. It is now signalled out of the near siding with a bogie bolster, a van and other stock. The combination of a blue diesel, a refurbished DMU, in the top right of the picture, traditional wagons and a steam era parachute water tank at the end of the platform, recalls the early post-steam era in which a traditional railway was operated by modern motive power. The explanation of the misleading headcode is probably that the lamps are being used as marker lights rather than as a headcode, in the transition period from headcodes or train numbers to marker lights for improved visibility on the front of all trains.

Above: The large urban freight depot worked on the same principles as the small yard, but unlike the country station, where passenger and freight facilities were on the same site, it was commonly a separate installation. Some, such as Curzon Street, in Birmingham, used the site of a former passenger station. Curzon Street, the original terminus of the Grand Junction and London & Birmingham Railways, was opened in 1837. The original station buildings, have survived. They were designed by Philip Hardwick, architect to the much lamented Doric arch at Euston. Passenger services were transferred to New Street station in 1854, from which time Curzon Street was used for goods. In a large goods depot, loading sidings were laid in pairs, with vehicle access between each pair of tracks, which was wide enough for road vehicles to stand beside each track and allow an adequate drive way for other lorries to arrive and depart.

Right: At the GWR's Moor Street station in Birmingham, there were goods facilities on two levels linked by a wagon hoist. Although the passenger station opened in 1909, it was not until 7th January 1914 that Moor Street goods opened, to relieve pressure on existing goods depots and provide a city centre terminal for certain classes of traffic. It did not, for example, cater for cattle or end-loading car or furniture van trucks. Birmingham Small Heath and Hockley had 20 and 25 ton cranes, but Moor Street only received a 6 ton crane, visible behind one of the ubiquitous English Electric 350hp diesel shunters, D3969. Moor Street closed to goods on 6th November 1972. The two vans show the peculiarities of the BR freight stock Diagram system. Diagram 1/208 vans existed with planked or ply bodies. Most had planked doors, but some planked vans had ply doors. They might have Morton brakes or 8 shoe clasp brakes. Given such diversity, one might wonder why it was that when a batch of vans with ply sides *and* doors was ordered, a new diagram was issued, but the right hand wagon, B772821 is a Diagram 1/213 van built as Ashford in 1956. By 1960 large orders for conventional vans had dried up, but in 1961 Pressed Steel produced a final 2,000 standard vans with ply sides and doors, basically to Diagram 1/213, but with Oleo pneumatic buffers. B786506 is to this final design, Diagram1/224. Note the different position of the chalking boards, the variations in placement of the lettering and different types of brake gear and axlebox.

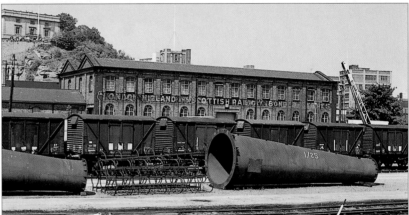

Below: The function of yards could change. The GWR Walsall Street goods yard, just south of Wolverhampton Low Level station, seen on 5th July 1976, was built to handle general freight and was reached from the GWR Birmingham to Wolverhampton main line. With the dismemberment of the proud GWR main line from Paddington to Wolverhampton and the Mersey, the yard became a steel terminal served by a connection to the former LNWR main line between Birmingham and Wolverhampton at Monmore Green. As at Curzon Street, the sidings are in pairs. The nearest road contains bogie bolsters and the second and third hold 4-wheel plate wagons. More bolsters occupy the fourth road, together with two trestle plate wagons, identifiable by the angled framework. These conveyed steel plate which was too wide to fit within the loading gauge if lying on the wagon floor.

Above: Few goods depots were more attractively located than this one at Nottingham, which was overlooked by the Castle, though I should emphasise that Nottingham Castle is more recent than the edifice of the Robin Hood legends. As well as providing transport, the railways offered warehousing and distribution and where goods were subject to customs duties, a bonded warehouse service as well. Although it is July 1976, the LONDON MIDLAND AND SCOTTISH RAILWAY BONDED STORES title, in white paint on a black background, survives on the brickwork of the old MR building. In the foreground are a rake of Shocvans, some with the older thin white stripes, others with the later short fat stripes. No two vehicles are identically lettered, whilst the van on the right has a continuous rainstrip on the roof instead of the normal three piece rainstrip. One Shocvan has the correct end shoc stripes. Others are so grimy that it is hard to tell and some appear to lack the end stripes completely. Adding interest are the massive tapered silo chutes in the foreground. If they were to be moved by rail, a bogie well or trolley set would be needed.

Above: At sorting sidings, to which traffic yards sent their wagons for marshalling into long distance services, public access was not required, so these yards consisted of rows of closely spaced parallel sidings, long enough to take a complete train or a portion of a train. Often located at junctions, where they received incoming freights to be broken down for remarshalling, they fulfilled a vital role. Banbury up yard was just north of the station, near the divergence of the GW main line and the connection from the Great Central main line at Culworth Junction, just south of Woodford Halse. A vast amount of freight flowed off the GC and was sorted at Banbury. We see the up yard at a quiet moment on 19th September 1965. In many places, as at Banbury, yards handled traffic in one direction, with the down yard in a different location. In some places, up and down yards existed side by side, but invariably functioned as two quite separate entities. A housing estate now occupies this site.

Right: The railways developed a range of ancillary businesses, and several companies became important shipping lines, and owned extensive docks too. Southampton Docks owed its success to the dynamic attitude of the London & South Western Railway, and later the Southern. The traditional freight berth was well supplied with travelling cranes, railway tracks and quayside warehouses. A multitude of sidings served the warehouses, throwing off connections on fantastically sharp curves. The Southern invested heavily in Southampton in the 1920s and 1930s, making it one of the most successful ports of its day. However the cost of traditional handling methods, with high levels of breakage and pilferage, prompted a switch to containerisation. By the time I took this view at Southampton on 26th July 1972, the era of the traditional rail served quay with its warehouse next to the berth was drawing to a close. For the modeller, it offers an interesting yet narrow layout with masses of shunting, and good scenic possibilities with shipping, cranes and warehouses which could be modelled in low relief, and the opportunity to use a wide range freight stock.

AN INDEPENDENT FREIGHT LINE

The formation of British Railways meant that eccentric little byways such as the Colonel Stephens empire vanished into the all-embracing nationalised system. Surprisingly there was one exception in Yorkshire, where the independent Derwent Valley Light Railway survived. The DVLR was a late-comer to the scene, its full 16 mile line between Layerthorpe station in York and Cliff Common opening to public traffic in 1913. At both ends, the DVLR made connection with the NER. It was promoted under the Light Railways Act 1896, one of the purposes of which was to encourage the spread of railways into rural areas In common with many light railways, the DVLR was excluded from the grouping in 1923, and remained independent. Passenger traffic had never been significant and ceased in 1926, but freight services continued, and it was only the closure of the BR Selby to Market Weighton line which led to the abandonment of the southern extremity of the DVLR in 1965. By 1972 just over 4 miles remained in use from York (Layerthorpe) to Dunnington.

Below: The headquarters of the DVLR was at York (Layerthorpe), where No 1, former BR Drewry 0-6-0 diesel mechanical, No D2298, arrives with a freight train comprised of 37 ton glw grain hoppers, on an unseasonable 16th July 1974. Just before the train arrived, the heavens opened and I was absolutely soaked.

Above: Under the DVLR's long serving and innovative manager, James Acklam, the company sought commercial tenants to bring traffic to the railway. At Layerthorpe land was made available for coal and oil depots and although the mileage on the DVLR was negligible, terminal charges and site rentals brought a useful income to the company, which continued to pay a dividend to its shareholders. This is the BP oil terminal at Layerthorpe in September 1979, the locomotive is the DVLR *Claude Thompson*, John Fowler & Co 4210142 of 1958, named after a long serving chairman of the company. The grimy state of rail tankers in the late 'seventies is apparent.

Top left: This hand worked 6-wheel breakdown crane at York (Layerthorpe) on 9th July 1974, the former BR DE902165, was built by Cowans Sheldon in 1892, whilst the match truck, 300445, was produced by the LMS at Wolverton in 1929. It was acquired by the DVLR, 'in case we need it', as James Acklam told me! At this time, the DVLR rolling stock comprised two former Drewry 0-6-0 diesel mechanical locos, 3 open wagons, a brake van, the crane and its match truck. After trials in 1976, the DVLR operated a tourist train in 1977-79, but the venture was not profitable and ceased. Real estate development eventually appeared more profitable than train operating. The coal traffic at Layerthorpe ended in 1980, and the Dunnington grain service in 1981, a sad ending to an independent line which had survived two wars, recessions, the grouping, and nationalisation.

Left: Coal and oil predominated at Layerthorpe, but at Dunnington, the far end of the line by the early 'seventies, agriculture was prominent. Thanks to the forward policies adopted by the DVLR, Yorkshire Grain Dryers established a plant at Dunnington. YGD acquired an 0-4-0 diesel mechanical shunter, John Fowler 4100005, *Churchill* of 1947, seen at Dunnington on 9th July 1974.

Bottom left: The DVLR was host to some of the last and most sophisticated vacuum braked wagons built. These were the celebrated 'Blues' of British Railway Traffic & Electric Co. BRT, a wagon hiring firm, introduced the first of their distinctive blue liveried 37 ton glw grain hoppers in 1965, building up to a fleet of 300 by 1971. An immaculate 7678 was one of the third batch, built by Powell Duffryn in 1967/68. It was vacuum braked, and fitted with Accelerator Freight Inshot when built. AFI was a braking distributor designed to improve the reaction times on vacuum brake trains, and reduce brake surges on long trains which could lead to broken couplings. BRT 7678, at Dunnington on 9th July 1974, represents the final flowering of vacuum brake technology before the triumph of the air brake.

STEAM AGE FREIGHT

As railways developed, identifying different train types became necessary. The simplest method was by a distinguishing headcode of lamps or discs on the engine. The earliest headcode indicators were round, so were known as discs, and when some companies used square indicators to give more options, they were called *square discs*! At first, each company used its own headcode, commonly based on train types, but sometimes on destinations. Gradually the train type system prevailed and a standard code evolved. The most notable exception was on South Eastern & Chatham, LBSCR and LSWR, where the density of suburban services, and sparse freight traffic, meant that signalmen were more interested in where trains were going, than the small differences in running speeds which the headcode system could indicate. For the moment, we will focus on the standard headcodes in the BR era.

Trains were split into ten basic categories, in a way which had developed piecemeal. As with many customs, though it was illogical, it was familiar, so change was resisted. On the LNER in the 1930s, trains were divided into Express Passenger, Ordinary Passenger, No 1 Braked Merchandise Train (composed of non-passenger coaching stock); No 2 Braked or Express Goods, Freight Trains class A, B and C, distinguished by average speed, a class D freight stopping at intermediate stations and a light engine or engine and brake. Instead of using average speeds as the determining factor, the LMS used the proportion of vacuum braked stock in the train.

BR adopted uniform bellcodes and descriptions, except where route discs prevailed. Where the LMS or LNER block codes differed, BR generally adopted the LNER bellcode. Trains were divided into 10 classes, A-K, omitting I. Table 1 in the Appendix shows the BR system as outlined in the 1960 General Appendix, with the standard block codes added. The first diesels did not carry alpha numeric route indicator boxes, but the benefits of such a system were obvious and in June 1962, the old classes A-K were replaced with 1 to 9 and 0 which followed after 9. The first character in the indicator box was a numeral, 2 for example indicating an ordinary passenger train. The second character, a letter, gave destination information and varied from area to area. S was used for Anglo-Scottish services for example. The remaining two characters were the reporting number within the train type and destination category, so a train might run as 6M43. The train categories were revised, and appear in Table 2, with bellcodes added. The main changes were that the old class C, which had consisted of parcels trains or empty coaching stock, both of which could run at passenger speeds, and fully fitted freights, was split up, the passenger stock services becoming class 3 and the fully fitted freights class 4. The 1962 regulations which refer to a train which is pipe fitted throughout with the automatic brake operative on not less than 90% of the vehicles, merits comment. Wagons could be hand braked, or vacuum braked in which case they were fitted with brake cylinders or they could be piped, but without vacuum brakes. The piped but unfitted wagon was intended to give greater flexibility in marshalling at minimum cost, as there were no complex vacuum fittings, merely a through pipe, Shunters had to be

careful not to include too many piped wagons in a train, for piped stock did not contribute to the braking capacity. Although the old class C had been divided into two classes, 3 and 4, the headlamp code remained unaltered! The splitting of one class meant something had to go to retain 10 categories and the victim was the hallowed 4-1 or coal train with its bottom left hand lamp code. Coal trains continued to run but were merged with the 1-4 unfitted through freight. Light engines or engine and brake, which had been class G in the old system, in the middle of the freight categories, were sensibly moved to the back of the list as class 0. This system lasted to the end of steam, and was also used by first generation diesels with headlamp discs and on those engines with 4 character indicators, in its numeric form.

Below: Pride of place amongst the BR freights went to the 3-1-1 or 'fully fitted' as it was known. The name was misleading, as the requirement was that all stock should be piped, but as late as 1960, the regulations provided that the automatic brake need only be operative on half the vehicles. In 1962, when the 3-1-1 moved from class C to class 4, the brake requirement went up to not less than 90%, with a 55 or 60 mph maximum speed limit. No 6911 *Holker Hall* passes Southam Road & Harbury in 1963 with an up class 4 freight. The stock, mostly vanfits, plus a few sheeted opens and Conflats, is typical of a 3-1-1 at this period. It does not contain a single wooden or steel bodied mineral wagon. The goods yard is reached via a trailing crossover from the down line in the distance by the shunt signals. In the up direction, there is a trailing crossover, partly visible in the foreground. A single slip provides a trailing connection from the up line into the yard, but not a facing connection off the down line.

Above: A young serviceman watches a Nottingham based Stanier 'Jubilee' No 45641 *Sandwich* on the head of a class 5 express freight at Oxford on 8th July 1964. The train will continue up the GW main line to Banbury, where it will divert on to the Culworth spur joining the GC and run via Leicester back to Nottingham. The 'Five', as it was also known to signalmen from its 5 beat bellcode, had a long history and unlike many trains, kept its character. As early as December 1924, it appeared in the LMS block regulations as an 'Express Freight or Cattle Train with the continuous brake in use on not less than half the vehicles'. This was too ambitious and under the 1934 block regulations the requirement was for the continuous brake to be in use on not less than one third of the vehicles. This continued as late as 1960 when the class D freight was one third fitted. Under the 1962 revisions it went to 50%, the figure that the LMS had decreed almost forty years before! Maximum speed was 50 mph. The second vehicle is an LMS Diagram D2039 van built during the Second World War.

Centre left: Stanier 8F, No 48171, heads south along the Trent Valley near Newbold on 27th July 1963 with a class 6 freight, or 1-2-2. The train is destined for Camden, as 48171 was a Willesden engine. Like the '5', the 1-2-2, with its fitted head of not less than four wagons, or a limited load, had a long history. Its LMS predecessor, running under the 2-2-3 bellcode, was the celebrated 'Maltese Goods'. At first class E under BR, the 4 wagon fitted head survived until the 1962 revisions, which laid down that not less than 20% of the train had to be fitted. On this occasion, there are 11 piped or fitted vehicles, to BR, LMS and SR designs, followed by 16 tonners, and some tank wagons. The 1-2-2 was restricted to a maximum of 45 mph. By now, the reason for this fine grading should be apparent, it encompassed a safe speed for the stock and the braking power available. Nine foot wheel base 16 tonners were not the kind of vehicles you could safely run at 60 mph, and even if you could, the problem with a freight train was not getting it started, but stopping it. With continuous brakes you could run faster as you had greater stopping power, without them, you ran slower.

Above: One of the imposing Gresley V2 2-6-2s, No 60966, leans into the curve at Chaloner's Whin Junction just south of York in May 1963, with a 1-2-2. A rake of Highs provides a fitted head and also act as barrier wagons to separate the steam engine from the petrol tankers which compose the bulk of the load. The silver or grey tanks indicate that these are class A 'highly inflammable' tankers, rather than

the less hazardous class B fuel oil tanks. Marshalling instructions required barrier wagons between the engine and any wagons carrying hazardous loads wherever possible and also between such wagons and the brake van. Where trains conveyed explosives and inflammable substances, barrier wagons were provided between such loads for obvious reasons !

Bottom: GC devotees would never forgive me if I did not feature the celebrated 'Windcutter' services which pounded up and down the GC main line until the mid 'sixties. An Annesley 9F, No 92088, enters Rugby station from the north on 8th July 1964, with an unfitted express freight or 3-2. Known as a class F freight until the 1962 revisions, it then became a class 7. The 3-2 was limited to 40 mph. The train comprises a couple of wooden bodied Highs, a steel High and several 12 ton Pipes identifiable by their double section drop sides and long 12 foot wheelbase. Further back are some 16 ton steel minerals. So much fitted stock at the head of an *unfitted* express freight seems illogical. The train could run as a 1-2-2 or even as a 5. If, however it was booked to run as a 3-2, it would be timed accordingly. The crew might have decided it was not worth the bother of connecting up the vacuum hoses and testing them if they were going to run to 3-2 timings anyway. Even if they had, they might have left the 3-2 lamp code, which the signalmen would be expecting. As we see here, a train could be made up of stock superior to its rank, but in the event of insufficient fitted stock being available, so that a train had to drop down the order, no latitude existed. If there were insufficient fitted wagons to provide the braked head for a 5, then the train would run as a 1-2-2.

Below: It is 30th October 1965 and the low winter sun is symbolic, for the sun is shortly to set on Western Region steam. No 48271 has taken water and blows off as it awaits 'the board' to head north via Banbury. Two Highs are followed by half a dozen Palbricks, which were a 1950s attempt to keep brick traffic on rail by providing wagons with removable sides, so that pallets of brick could be loaded by forklift truck. End tensioners kept the load in place and minimised breakage. Despite such revolutionary ideas, and 1420 wagons, the decline in brick traffic continued, and by the late 'sixties, the sight of a Palbrick in use was rare.

Top left: The 1962 revisions merging the class H 'through freight' and the class J 'mineral or empty wagon train' into a new class 8 through freight, using the head lamp and block code of the class H, drew down the curtain on the venerable 4-1 bellcode and single lamp above the left hand buffer (as you observed the engine) of the traditional coal train. Ever since its inception, the Midland main line from Leicester to London had seen a succession of coal trains trundling south from the East Midlands, Notts and South Yorkshire to feed the insatiable demand of the metropolis. Outside framed Kirtley 0-6-0s had crawled along in pairs averaging 8 to 9 mph. Johnson 0-6-0s had come and gone and in the 1920s the awesome LMS Garratts arrived. In the 1950s the finest freight engines of all did battle on the line, the celebrated Riddles 9Fs. Many trains changed engines at Wellingborough, here a grimy No 92126 clanks south through St Albans on the up slow line on 23rd May 1959.

Centre left: 'The Coal Train is Dead; Long Live the Coal Train.' Carrying the new class 8 headcode of 'one top, one centre', a Frodingham-based Riddles Austerity 2-8-0, No 90166, erupts into volcanic action as it bypasses Chesterfield station on the up goods line with a coal drag on 24th July 1964. The leading wagon is an early riveted 16 tonner without the strengthening lip at the top of the body turned over to form a channel section as in later batches. To the enthusiast, the rich exhaust was enchanting, but not perhaps to the housewife who lived near the line and for whom it was washday !

Right: The 1-4 was restricted to 35 mph and could convey unusual loads which were not permitted on faster services. Riddles Austerity No 90471, an Eastern Region engine from Canklow, eases a southbound class 8 freight on to the top of the 1 in 37 Lickey incline at Blackwell on 31st July 1964. The leading wagon, which, with its fishbelly girder, a derivative of the Warflat, serves as a match truck for the second vehicle, a Lowmac which is conveying a large item of contractors plant, which overhangs the end of the wagon. Further back is a mixture of 16 tonners and general stock.

Above: Wayside stations were shunted by class K, later class 9, freights, identifiable by one headlamp above the right hand buffer, as you looked towards the engine. Stanier 8F No 48325 has climbed out of Leamington, but has steam to spare as she ambles through Southam Road & Harbury station with a class 9 freight on 8th August 1964. The goods yard had closed on 11th November 1963 and the cross over in the foreground is out of use. There is a wooden chock between the blades and the right hand running rail, and the left hand blades are clamped. The train is a typical pick-up goods. The leading wagon is a 30 ton Bogie Bolster C, to Diagram 1/475 followed by a set of 4-wheel Insulfish vans. The BR designed vans could be distinguished from their LNER predecessors by the presence of heavy diagonal angle iron on the ends, as here. Next comes a Conflat with a type A container, some class A petrol tanks, identifiable by their red solebars and grey tank bodies, and other oddments.

Bottom: The 'Light Engine' and 'Engine and Brake' shared class G, later 0, and a headcode, one lamp at the centre of the buffer beam, but had separate bellcodes, 2-3 for a light engine and 1-1-3 for an engine and brake. The 56xx 0-6-2T was developed from pre-grouping designs by Collett for the heavily graded South Wales valleys lines. No one could call the 56xx beautiful, but its slogging ability made it ideal for short haul freights. Some 56xx tanks appeared at the Wolverhampton division sheds of the GWR, later BR area 84. No 6644 has arrived from Banbury to shunt the APCM works at Greaves Sidings between Fenny Compton and Leamington on 4th July 1964. To the left are a pair of GWR 'toad' brake vans, whilst the brake with No 6644 is to an LMS design. The two iron ore hoppers in the sidings, B435710 and B435694, were built to Diagram 161 by Charles Roberts & Co in 1949 and originally rated at 22 ton, later upgraded to 24 ton. Despite their designation, they are being used for chalk traffic for the Blue Circle works at Greaves.

Above: The Southern Region, with its dense network of routes around London, operated differently, using discs or lamps to give route information, rather than indicate train types. As well as the normal 4 lamp irons, three on the buffer beam and one below the chimney, the Southern used brackets each side of the smokebox. Using one, two and three disc codes, with 6 brackets to built up permutations, the Southern Region was able to provide thirty different options, compared to the ten used elsewhere. The code on No 73016, at Basingstoke on 29th October 1965, indicates that it was working on the former LSWR line from Waterloo and Nine Elms to Southampton.

Below: To cover every code on the Southern would be impracticable, so Southern devotees will need to refer to the standard works on SR route disc codes, but we will look at this remarkably clean Standard class 4MT No 2-6-0, No 76008, at the west end of Salisbury station on 19th May 1964 with bogie ballast hoppers. The code, one up, one down, is for the Waterloo-Salisbury-Plymouth section. The locomotive was a Salisbury engine from the start of the 'sixties until a month before its withdrawal in May 1967, when it was transferred to Bournemouth. To see a clean Standard at this date was rare.

TRANSITION ERA FREIGHT

The Train Classification regulations appeared in the General Appendix to the Working Time Table. The 1968 supplement to the Appendix retained the traditional categories, but added a new class 3 train, the Freightliner, which was limited to 75 mph. A more comprehensive revision took place in 1969. Class 3, parcels, empty coaching stock or Freightliner trains was split into three new classes. First was the new class 3, or an express parcels train consisting of stock permitted to run at 90 mph or over. Freightliners moved down to class 4, which they shared with Parcels stock permitted to run at 75-90 mph, whilst ECS became class 5. The three classes of fitted freight remained, but moved down the order, becoming classes 6-8. To keep within the required 10 classes, the unfitted freight was trimmed from the three categories of the 1962 revision, 3-2, 1-4 and 3 to a single class 9 'unfitted freight train', whilst light engines, or an engine and brake remained class 0. These rules, which were repeated in the 1971 revision and the new 1972 General Appendix provided the basis under which hand and vacuum braked stock were to operate for much of their twilight years.

Below left: The earliest diesel locomotives, such as LMS 10000, or the English Electric 350hp 0-6-0 shunters, had steam style lamp brackets, and conventional headlamp codes. D3106, one of six shunters (D3105-3110) allocated to Banbury, features in this rare *on the road* scene. The 08s, as they later became, had a top speed of only 20 mph, so were not as versatile as the 0-6-0PTs which they replaced, and seldom appeared on train duties. The solitary headlamp proclaims this is a class 9 stopping freight. The LNE/BR standard brake at the front, a rake of steel or timber Highs conveying scrap sleepers, and a GWR Toad brake at the rear, indicates this is civil engineer's working. The date is 4th October 1964, and the location is just south of Banbury, the line in the left distance being the now closed LNWR branch from Verney Junction to Banbury.

Below right: The prototype Southern main line diesels, Nos 10201-10203 used electric headlamps by night, and metal discs by day, which folded over on themselves when not in use, and were opened upwards to provide route indications. Despite the predominance of LMS officers on BR, this was adopted for the first production diesels. D309 in original green livery, trundles through Woburn Sands on the Bletchley-Bedford line with a short class 7 'Express Freight' or 3-2 on 7th April 1967. The upper and centre discs are offset to the left, because of the nose doors. Except for the small yellow warning panel, the engine is as built, one of the original batch of EE Type 4s, D201-323 with train discs. By the time D324-345 came out, the idea of an alpha-numeric train describer had taken hold and a divided display panel was provided, with two characters each side of the nose doors. By 1961 when the final batch, D346-399 appeared, the doors were eliminated, in their place was a central 4 character display. The train is short, a single goods van, some 16 tonners and a brake.

Right: A beautifully clean class D5688, one of the production series class 31 Brush type 2 A1A-A1As, eases through the former MR Lincoln St Marks station in October 1965. Although the pilot scheme Type 2s lacked route boxes, the production engines carried them and looked very smart in their striking livery of standard Brunswick green with two white bands and a light grey roof. At this time, Regent were taking delivery of their first 45 ton glw class A petrol tankers and the wagons looked equally smart. Such a combination, a clean engine and wagons, was seldom encountered other than in 'Modernisation Plan progress' publicity photos. The code, 7M07, reveals that the train is another class 7 unfitted express freight, and that it is destined for the LM Region.

Above: A Derby Type 4, No 6 *Whernside* heads a class 9 unfitted freight along the Erewash Valley line near the 840 yard long Alfreton tunnel on 19th March 1971. Compared to steam days, there are several changes. Although most of the train consists of unfitted steel minerals or hoppers carrying coal, the first four wagons are 45 ton glw tankers, the requirement of barrier wagons between the locomotive and petroleum product tankers having been abolished. Until the 1962 amendments, when the '1-2-2' moved from a minimum of 4 fitted wagons to 20% of the vehicles in the train, this fitted head would have brought the train within one of the express freight categories, but by 1971, even if the smaller fifth tank was vacuum braked (which is unlikely), the fitted head is insufficient to give the train fitted status. Only ten of the original 'Peaks' were built, the design being overtaken by improvements, and by the mid-'seventies, their days were numbered. After a spell at Camden for the West Coast Main Line, they moved to the Midland division, operating predominantly on freight duties out of Toton, prior to their withdrawal between 1976 and 1980.

Above: A pair of English Electric Type 1s, later class 20s, 8014 and 8016, carrying the class 9 unfitted freight code, squeal round the sharp curve from Syston East Junction to Syston North Junction with loaded ironstone tipplers from the BSC quarries at Corby, on 7th March 1973. Despite the capabilities of the motive power, whilst traditional hand braked 4-wheel wagons predominated, train speeds remained low, and when drivers did turn on the power too enthusiastically, there was a spate of derailments. Extensive research by BR showed that springing and tyre profile defects could cause rhythmic hunting patterns at critical speeds. Some remedial work was done, but the trouble strengthened the desire to eliminate the traditional wagon. The headcode discs, which had replaced traditional headlamps, were rare by 1973, four character headcode boxes having been fitted to many locomotives by this time.

Top: The glass industry is a major user of high grade silica sand, which, with the concentration of glass making in a few areas, often travels considerable distances. Pilkington's works at Ravenhead near St Helens makes a range of specialist glass. In the 1970s silica sand was conveyed in open hopper wagons. E3072 powers north alongside the Oxford Canal at Brinklow on the Trent Valley line on 5th October 1971 with the Ravenhead sand train, 8M17. Originally the hoppers were unfitted, so vacuum fitted 16 ton steel minerals were marshalled to bring the train up to partially fitted class 8 standards. Once some hoppers were vacuum fitted and repainted into bauxite livery, the need for empty 16 tonners for braking purposes diminished. By October 1971, vacuum braked hoppers outnumbered the Minfits in the fitted head, but the need to run empty mineral wagons merely to provide braking was a reflection on the muddled wagon policies of the previous 20 years.

Above: Integral electric headlamps with headcode discs for daytime use were the way forward in the 1950s, replacing the old fashioned oil headlamp. In the 1960s, the 4 character train code box was the answer. By the mid-'seventies it too was a victim of modernisation, for with an increasing amount of the network worked by power boxes, with train codes sent from box to box automatically, the need for trains to carry any visual identity was diminishing, as signalmen might never see the trains passing through their territory. The use of the indicator boxes ended, but until they were blanked out, a variety of codes might appear, such as 0000 as in this case. Sometimes drivers displayed the engine number or facetious codes. BR Standard Type 2, No 24 133 enters the up goods loop at Cosford, on the Wolverhampton-Shrewsbury line on 1st July 1976. The train is typical of wagon load freights of this period with an amazing mixture of stock.

WHEN COAL WAS KING

Coal was the mainspring behind the evolution of railways in the North East, and the earliest railway vehicles were coal wagons. The first steam locomotives worked in the collieries and coal remained the most important traffic throughout the steam age. In 1938 there were 583,789 private owner wagons operating on the mainland railways as against 660,155 railway owned vehicles. Most of the POWs were for coal or coke. Although most coal was moved in POWs, the railways had their own mineral wagons as well, so approximately half the wagon stock was used for minerals. At the end of 1967, BR had 458,475 wagons, of which over 280,000 were for minerals. The drop from 580,000 to 280,000 seems enormous, but many of the 1938 wagons were 10/12 ton wooden bodied designs. By 1967, the majority were 16 tonners, with a few 21 or 24.5 tonners. Magnificent though the A4s, Duchesses, Lord Nelsons or Kings were, what really mattered, in financial terms, were nondescript goods engines hauling nondescript mineral wagons.

Above: One of the most colourful aspects of the pre-1948 scene was the host of privately owned wagons. On the outbreak of the Second World War, POWs, other than specially constructed vehicles (such as tipplers or tank wagons), were requisitioned by the government, to permit more intensive working. They were never returned to their owners, and passed to BR in 1948. The scarcity of pre-war colour material means that good views of POWs are rare. This shot, at Cambridge in May 1942 of LNER K3 No 2447, arriving with a loaded coal train shows the condition of many POWs under government wartime neglect. The leading wagon, with its bright red livery and large white lettering shaded in black, might almost have been selected to depict one of the most popular PO liveries. It is a Bullcroft Colliery wagon, but with pooling of wagons, is unlikely to have come from its home pit, which was located between Carcroft and Adwick-le-Street on the West Riding & Grimsby Joint (GC & GN) in Yorkshire.

The second wagon belongs to Dalton Main Colliery Co. Again, the fates have been kind to us as this displays the curved title used by many collieries to make their wagons stand out. Dalton operated two pits in Yorkshire, Roundwood and Silverwood. The third wagon, Wharncliffe Woodmoor, displays another common style with the lengthy title in smaller lettering than Bullcroft. Its location is given on the second line, lower down. Wharncliffe Woodmoor was on the former GCR or H&BR/MR near Cudworth. The next wagon displays another distinctive style, with its large 'SC'. This stands for Stephenson Clarke & Associated Companies Ltd, which operated over 10,000 wagons, which could be seen throughout England and Wales even before pooling. Liveries included dark grey or light grey with indian red corner plates, and black or red oxide. Most carried a prominent SC on the side, but in some cases this was omitted. *Colour-Rail NE29*

Left: This study of Littleton Colliery internal user wagon No 16, taken on 22nd May 1976, provides a closer look at a typical POW. At first glance, it could have been built at any time from the 1880s to the 1930s, but closer inspection reveals it is to the RCH 1907 specifications with updating. It has semi-circular crown plates on the solebars above the axleguards which were typical of the 1907 wagons, but unusual after 1923. The small metal brackets which project above the headstock were to restrain longitudinal movement by the body and were another 1907 feature which was dropped after 1923. On the other hand, the ribbed buffer guides are more common on post 1923 wagons, as is the label clip on the solebar, rather than on the curb rail. The wagon was probably built with single side brakes, but modified to two sets of independent brakes to meet new RCH requirements in the 1930s. Now in NCB internal user colours, with the prominent white cross, traces of its former red oxide livery can be seen, along with fragments of white lettering shaded in black, but not sufficient to be certain of its previous ownership. *Chris Nettleship*

Below: Before seeing how the traditional wooden-bodied private owner wagon gave way to the 16 ton steel mineral, we will look at the collieries. Coal is associated with Scotland, the North East, Yorkshire, Lancashire, the Midlands and South Wales, but many people forget there was once a Kent field, which covered some 80 square miles north west of Dover. Unlike the better known fields, it did not come into production until the early years of the twentieth century and as with earlier fields, its development prompted railway building. The railway most associated with this new coal field was the East Kent Railway, a part of Colonel Stephens' quaint empire, which gave the whole project an aura of whimsy. The EKR had its own station at Shepherd's Well, alongside the SECR Dover to Canterbury line. It opened to public traffic to Wingham in 1916, serving collieries at Eythorne, Tilmanstone, Woodnesborough and Wingham. We are looking north from the SECR Shepherd's Well signal box across the exchange sidings in July 1973. The two ex-GWR Toad brake vans provided a fitted brake head on the Tilmanstone colliery trips. Next to the brake vans one of the two 16 tonners is of the earlier riveted construction with a simple angle at the top, the other is the more common strengthened version with turned over channel section lip around the top of the body.

Below right: Sadly the Kent coalfield did not achieve lasting success and after Wingham closed in the 1930s, the East Kent Railway was dependent on Tilmanstone, 2 miles north east of Shepherd's Well. Passenger services ceased in 1948, and general freight in 1951, leaving only the section to Tilmanstone open for coal traffic. We are looking towards the winding gear at Tilmanstone in July 1973. The mix of brick and corrugated iron buildings typified the coalfields, as do the 16 tonners. The right hand vehicle, B261424, is one of 200 Diagram 1/108 wagons built by Butterley in 1958, part of the penultimate delivery of these sixteen tonners, the most numerous type of wagon ever to operate on British Railways. The left hand vehicle, B558588, came from the final Lot No 3146 for 2,000 Diagram 1/117 wagons and also dated from 1958.

Bottom right: The NCB acquired many surplus 16 tonners from BR to replace its elderly wooden bodied internal user wagons. They were repainted in various liveries, including black, red and light grey. If wagons lost their BR plates during refurbishment, identification was difficult. These wagons at Tilmanstone in July 1973, with large 'Internal User Only' crosses, record some varieties in 16 ton design. Both are riveted with top flap doors, and may appear identical. However, No 30 has double brakes, whilst 31 has the Morton brake. No 30 has a single triangular gusset supporting the body above each axleguard, whilst 31 has a pair of I shaped gussets. The triangular gusset was characteristic of the earlier 16 tonners ordered by the Ministry of War Transport and later by the Ministry of Transport whilst the double I gusset was common on Diagram 1/109 wagons ordered by BR. Unless body transplants have taken place, which is unlikely with riveted stock, 30 is to Diagram 1/105, with top, side, bottom and end doors and was ordered by the MOT, but not actually delivered until 1948-49, whilst 31 is to Diagram1/109, without bottom doors, the type being built between 1950 and 1957.

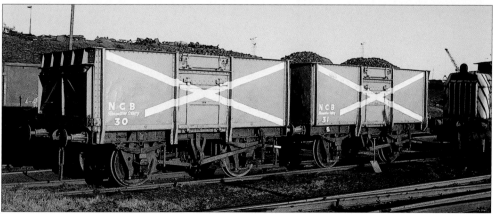

Above: Although the 10-12 ton wooden POW dominated the coal trade until BR days, there were exceptions. Under G J Churchward's relentless probing for efficiency, the GWR built 20 ton steel bodied coal wagons from as early as 1901. Their lower tare weight per ton of load, and the reduced siding space they occupied was not lost to GW management when their association with the South Wales coal trade dramatically increased with the absorption of the Welsh valleys lines at the grouping. The GWR espoused the cause of the 20 tonner with a messianic zeal and thousands of standard 20 tonners were leased to the South Wales coal owners. Some carried the colliery's own lettering. Others were kept on a dedicated circuit working with prominent 'EMPTY TO' instructions. No 63066 was built in 1946 to Diagram N34, and is now preserved at Didcot. This 1940s design is little changed from the original 1901-1903 Diagrams N1-N3.

Centre left: Some private owners, such as Richard Thomas & Co or the West Midlands Joint Electricity Authority, bought 20 tonners, a few of which survived alongside their more modern BR counterparts in the 1970s. Upgraded to 21 tons during the war, former private owner P152688K is at Syston Junction in June 1973. The K suffix was applied to many 21 tonners to distinguish them from 16 or 24.5 tonners to assist loaders and guards in their duties. This view, from the now demolished MR Syston North Junction signal box, shows the train in the foreground, moving left to right, and in the distance, its locomotive, 'Peak' D10 *Tryfan*, which has turned through almost 180 degrees around the sharp chord connecting the MR main line and Syston to Peterborough route, heading right to left. The signal box to the left of D10 is Syston East Junction. The wagons to the right, B315748 and B316848 were from a batch of 2,600 all-welded 21-ton replacement bodies built in 1971/72 at Shildon on underframes from earlier BR 21 or 24.5 tonners. The similarities with their much older neighbour are noteworthy.

Above: The Ideal Stocks Committee shared the GWR liking for higher capacity wagons, but instead of selecting the proven 21 tonner, preferred the taller 24.5 tonner. As a result, production of the 21 tonner was modest, with less than 7,500 wagons being built between 1950 and 1963 when construction ceased. B312677K, at Coundon Road coal yard, on the Coventry to Nuneaton line, on 3rd June 1971, was built at Derby to Diagram 1/120 under Lot 3430 in 1962/63. Unlike earlier 21 tonners, it was fitted with 8-shoe self adjusting clasp brakes, vacuum equipment and roller bearings.

Top right: In the quest for greater capacity, the Ideal Stocks Committee recommended the 24.5 tonner. B280766 was built to Diagram 1/115 under Lot 2602 at Shildon in 1954, is at Syston North Junction in June 1973, by which time it was branded COAL 24. The vehicle to its left is another 24.5 tonner. Unlike the standard 16 ton steel mineral or 21 tonner, which were the same height as the older wooden 10/12 tonners, for which the colliery screens had been designed, the larger wagons were too tall for many collieries, so their use was limited. These limitations meant that despite endorsement by the Committee, less than 4,000 were built. By 1982 all had been withdrawn. Had BR followed the Great Western lead, and standardised on the 21 tonner, the results would have been better.

Right: Although the Ideal Stocks Committee advocated the 24.5 tonner, they recognised this might present short term problems, so reluctantly proposed the 16 tonner as an interim measure. Many thousands had been built for the LMS, LNER and the Ministry of War Transport, but in the fifties about a quarter of a million 16 tonners poured out of BR and contractors' workshops in the biggest wagon building spree in British history. Over 200,000 were built to one design, Diagram 1/108, between 1950 and 1958. B99038, a Diagram 1/108 wagon built by Pressed Steel in 1950 to Lot 2259, seen at Rugby on 28th June 1971, recalls the livery changes introduced in 1963. The designation, COAL 16, load, tare and number are carried in a framed box on the left hand side, and subsidiary data in very small lettering in a small panel at the right hand end. Although most 16 tonners were built with drop-flap top doors these were little used by coal merchants and few photographs show the doors open. There were many detail variations, not least in brake gear, which ranged from unfitted stock with Morton brakes to clasp braked fully-fitted wagons. B99038 has Morton brakes, with the dog clutch on the same side as the brake gear. The brake lever is ribbed. The chassis, brake gear and running gear are black, but have picked up rust and dirt, so the one colour not in evidence is the glossy black beloved of modellers. The only black areas are the right hand axlebox and the dog clutch which have had an application of oil.

Bottom right: Whilst wagons to Diagram 1/108 were pouring out of the workshops, BR introduced another design, Diagram 1/117. Given the constructional differences within Diagram 1/108, it is hard to see why! Branded as COAL 16 VB under the 1963 livery, B551778, at Fakenham on 11th July 1971, was built under Lot 3145 by Pressed Steel, as part of a 1957 order, but plated as delivered in 1958. Out of the 8,250 wagons built to Diagram 1/117, three lots, totalling 2,700 wagons, were vacuum braked from new. The open door provides a rare opportunity to see the internal livery of coal-grimed rust.

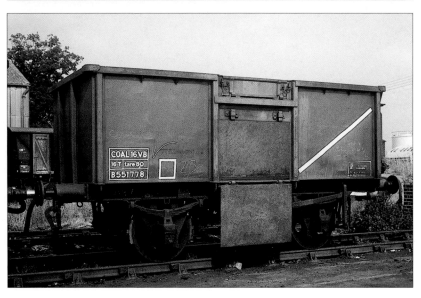

Above left: By 1974, the days of wagon load coal being delivered to local stations throughout the country were drawing to a close. B115822, a Diagram 1/108 steel mineral, built by Cravens in 1952, is at Llandeilo on 29th June 1974 in a scene which might have been posed to demonstrate traditional methods. Both side doors are open, the nearside door being propped, a practice favoured by coal merchants, but unpopular with railwaymen due to the risk of accidents if the door was not adequately supported. Coal has been shovelled out to lie each side of the wagon, and a start has been made on bagging. To the right of the wagon is the merchant's coal scale and by the lean-to shed are wicker baskets more bags and other impedimenta. The sheds are dilapidated, to say the least. In 1968, wagon load freight had constituted over two thirds of freight tonnage. By 1972 it had dropped to a third and by 1976 to a fifth.

Above right: Until the miners' strike of 1984, the 16 tonner, although declining in numbers, was still common, but the damage done to many wagons which sat under load for weeks, plus the draconian cut-back in the coal industry after the defeat of the miners, saw the 16 tonner go into rapid decline. Many were transferred to Departmental use, such as DB554826, which was a vacuum braked example built to Diagram 1/108 under Lot 2912 by Cravens in 1957. Often used for spent ballast, the 16 tonners' high cubic capacity offered pitfalls to the unwary, for substantial overloading was possible, given the density of spent ballast. As the channel section at the top provided vital structural strength, it was not possible to cut down the sides without a costly rebuild, so overloading was tackled by burning rectangular holes in the sides. In theory B554826, as a fitted vehicle, should be painted bauxite, but the main colouring is rust.

Below: The traditional British method of coal handling was archaic even in 1900, yet there was a better answer, the hopper wagon which could discharge its load in seconds. The Great Western made limited use of hoppers and the steel bodied 20 ton Diagram N12 coal hoppers of 1905 were some of the most advanced wagons of that era. A similar batch, built to Diagram N25, followed between 1923 and 1927. They were used to convey gas coal from Wath Main Colliery, Barnsley, to the GWR gas works at Swindon and for supplies to Park Royal power station in London. A pair of 20 ton hoppers are being shunted, together with an extremely grimy No 5974 *Wallsworth Hall*, at Swindon on 5th September 1937. The nearer wagon, 53071, displays the pre-1936 livery, with GW in large letters whilst 53081 carries the 1936 livery with small GW and numeral. *Perrier / Colour-Rail GW 76*

Top right; Although sticking to timber construction, the leading exponent of hopper operation was the North Eastern Railway which pioneered the large scale use of 20 ton wagons and elevated sidings running over coal drops. Instead of hours of back breaking effort shovelling coal, the hopper door was opened and in a few seconds, 20 tons of coal had been gravity fed to a stockpile beneath the siding. This preserved example, an NER 20 ton hopper, seen at Shildon on 29th August 1975, recalls this logical way to handle coal. The reason it did not spread included parochialism amongst railway companies, a reluctance to spend money on costly coal drops, the fragmented nature of the coal trade with thousands of small merchants who were happier with what they knew and the fragile nature of the coal from some fields. *Mike Hodgson*

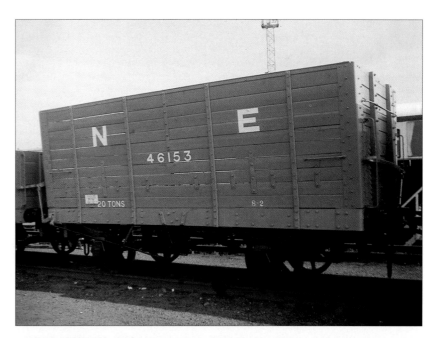

Centre right: After adding over 7,000 wooden hoppers to the NER design, the LNER switched to a steel 20 ton hopper in 1936, over 8,000 were running by 1948. E307163 is at Swanbourne Sidings on the Bletchley to Oxford line on 4th March 1967. The brake lever on this LNER designed wagon recalls the NER practice of extending the handle well above solebar height on hopper wagons.

Below: Because of its localised nature, the Ideal Stocks Committee largely ignored hopper coal operations, so missing an opportunity to modernise the trade. However, several thousand BR hoppers were built, with the capacity revised to 21 tons. At first they had the unusual NER/LNER brake lever seen in our previous illustration, but a conventional brake lever was fitted to later batches. B429911, designated HOP 21 VB, at Shepherd's Well in 1973, is to Diagram 1/146, and came from Gloucester Carriage & Wagon under Lot 3159 in 1958, by which time construction of conventional coal wagons was coming to an end. It has been upgraded with roller bearings and vacuum brakes. The livery appears to be a mix of rust and bauxite.

IRON ORE, STONE, SAND AND CEMENT

After coal, one of the most important mineral traffics was ironstone. Iron ores occur in various parts of Britain, most notably west Cumbria, where high grade deposits of haematite were an important source of iron for many years, but also in Cleveland, the Black Country and in a band stretching across the south and east Midlands, with workable deposits in Warwickshire, Oxfordshire, Northamptonshire and Leicestershire. Ore was also imported. Today virtually all Britain's ore needs are met from imports, rather than home extraction. The railways had to carry the ironstone from the ore fields or ports to the steel works. Sometimes these were adjacent, as at Corby, where a modern steel plant was erected adjacent to large quarries, but when local measures were exhausted, as in the Black Country, ore came from further afield. Iron ore is much heavier than coal and because of this ore hoppers were much smaller than coal wagons. An ordinary coal wagon, if filled with iron ore, would be dangerously overloaded.

As the steelworks required trainloads of ore, shovelling was unacceptable and quick and economical discharging was a must. Hopper wagons were used from Victorian times, and in 1939-40, Stewarts & Lloyds commenced the first large scale use of rotary tipplers, in which the wagon was turned upside down in a tippler frame to discharge its load. Although BR was to build hoppers of its own, after nationalisation the emphasis was on tipplers. As limestone, chalk and other minerals are denser than coal, the types of wagon which had evolved for ironstone soon found their way on to these traffics too.

Above: Although dating from 1948, this portrait of LMS Garratt No 7987 at Copmanthorpe with empty ironstone hoppers from Skinningrove to Desborough, is pure LMS. The 'up and down' lamp code indicates the train is a 'through freight' under the 1-4 bellcode. The second and sixth wagons are 20 ton hoppers to Diagram D1669, still in pre-1936 livery, with large LMS lettering. *Sanderson/Colour Rail LM 90*

Below: A Standard class 4 tank is not the motive power one associates with mineral trains, but a relatively clean No 80072 pulls away from the down refuge loop at Fenny Compton on 27th July 1964 at the head of a rake of iron ore hoppers conveying chalk for the Blue Circle Cement works at Greaves Sidings between Fenny Compton and Southam Road & Harbury. By the whitened state of the frames, these hoppers, although still badged Iron Ore, have carried chalk for a considerable period. No 80072 had been used on suburban services on the former London, Tilbury & Southend section before migrating to Leamington, whose 2L code is displayed on the smokebox, the code being adopted after the transfer of this former GW line to the LM Region.

Right: Iron ore hoppers were also built for the large iron and steel companies and for the Ministry of War Transport. P101453 was built by Metropolitan-Cammell to a Charles Roberts design in 1940 and was taken into BR ownership in 1948. In the early 1960s, several ore hoppers were transferred to the Eastern Region Civil Engineer to carry slag ballast from steelworks. They received a D prefix to the number, and a shallow extension to increase their cubic carrying capacity. In 1978 DP101453 was transferred to the Isle of Wight, but was unpopular, for unlike proper ballast hoppers, it was only provided with a centre discharge door. When photographed at Sandown in July 1994, it was condemned. Although well rusted, it still carries the gulf red paint which would have been applied when it was transferred to Departmental stock in 1962.

Centre: Between 1949 and 1959, BR acquired 5,270 ironstone hoppers to eight separate diagrams, which varied in capacity from 22 to 33.5 tons, but because iron ore is much denser than coal, they were quite small wagons. B436166 was built to Diagram 162 by Birmingham Carriage & Wagon Co in 1950 and was an uprated 24 ton version of the 22 ton Diagram 161 wagon. The latter were the first large batch of ore wagons built for BR. Although designated as IRON ORE hoppers when built, they soon found other uses. Limestone was another heavy mineral, with up to twice the density of coal, and although still branded IRON ORE, these hoppers are on a limestone working at Derby in 1976

Below: Sand is another heavy commodity and by the early 1970s many iron ore hoppers had been redesignated SAND. E3075, a class AL5, later class 85, heads train 8M17 north through Rugby on 28th June 1971. The working is the 10.33am from Redhill to Ravenhead Junction, conveying high quality silicon sand for Pilkington's glass works. The leading wagon, B438580, is a Diagram 1/163 iron ore hopper built at Shildon in 1955 refurbished with vacuum brakes to provide a fitted brake head on the heavy Pilkington sand trains. It has been redesignated SAND VB.

Left: This detailed portrait of B438581 at Rugby on 16th March 1971, shows a Diagram 1/163 iron ore hopper, built at Shildon in 1955, in virtually original condition, still with hand brakes but designated SAND. One of the problems with moving sand at high speeds in open hoppers such as this was that a good deal could be blown out in transit. This was particularly unpleasant for guards, who protested over this form of sand blasting. After numerous complaints, a van was marshalled as a barrier wagon between the last hopper and the brake van. In this instance, the van acting as the barrier wagon is B853241, a Diagram 1/209 Shocvan. The way in which sand has lodged on the ribs of the corrugated end of the van proves that the guards had a point and a barrier wagon to protect them from the sand storm created by the train's progress was very necessary.

Left: A Great Eastern branch once ran east from King's Lynn to Swaffham and Dereham. All except the last three miles to Middleton has long gone. The remaining section serves the British Industrial Sands Ltd quarry at Middleton Towers. BIS No 46, otherwise loco No 4, a Hibberd Planet diesel (FH 3910 of 1959), shunts a rake of five loaded hoppers in 1975. The nearest wagon, B437546, is a 33.5 ton Diagram 1/167 iron ore hopper built under Lot 3002 at Shildon in 1957, but now redesignated for sand.

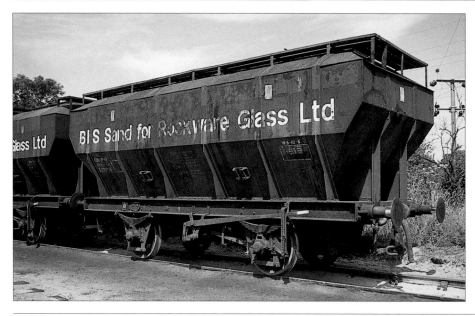

Left: The Covhop was one of those vehicles which did not sit easily in any classification. When it first appeared in 1952, it was to Diagram 1/192, following on after iron ore hoppers, but BR later reallocated them to page 210 in the Diagram book amongst conventional goods vans. Until 1958, Covhops were hand braked, but four piped lots followed by 1961. The final Lot 3431 of 1962 came from Ashford, and comprised 130 wagons or almost 10% of the total production. These were equipped with vacuum operated disc brakes, then a revolutionary new idea. B870726 was from this final batch, and is at British Industrial Sands, Middleton in 1975. The dedicated traffic branding on a BR-owned wagon became increasingly popular in the 1960s and 1970s.

Right: Although the volume of cement traffic never approached that of coal, it was, and still is, an important source of revenue. Cement is produced by mixing clay and chalk in water to produce a slurry, which is then pumped into a rotary kiln where it is heated to drive off moisture and produce a clinker which is ground to a fine powder. Before the days of dump trucks, quarries often had their own railway systems. As the quarries became deeper, grades became steeper until even 4 or 5 wagon trains became a challenge for the industrial tanks which blasted up and down the inclines. Associated Portland Cement, better known as Blue Circle, was a major operator with quarries and works throughout the British Isles. APCM No 5, a Hunslet of 1929, is propelling empty chalk wagons down to the working floor of Swanscombe quarry in Kent on 6th January 1968. The wagons were internal user, and hard use has resulted in their sides being severely bowed. They were emptied on a side tippler, so have no side or end doors. Few materials leave a more characteristic or spectacular weathering than chalk and railway wagons used on chalk trains became predominantly white. *Colour-Rail*

Below: Rugby Portland Cement developed from owning one plant at Rugby to operations in a number of countries. One RPC plant was the former Kaye & Company's lime and cement works at Southam, on the single track LNWR branch from Weedon to Marton Junction, where it joined the LNWR Rugby to Leamington line. The Weedon to Marton Junction line closed to passengers in 1958, but chalk trains from Leighton Buzzard continued to arrive at Southam cement works via Weedon until the line was closed as a through route on 3rd December 1962, after which chalk trains ran via Rugby, reversing at Marton Junction, the Marton Junction to Southam line being worked on the one engine in steam principle.

Usually powered by Stanier 2-8-0s, the Rugby to Southam chalk trains were some of the last steam workings in the area. We are looking from the road overbridge by Southam station towards the RPC works in June 1966. An 8F, No 48534, is shunting in the distance.

Left: The other approach to moving minerals to bulk users was to adopt the rotary tippler. Stewarts & Lloyds selected the 14 ton Charles Roberts slope-sided steel mineral wagon but without side doors for the first large scale use of rotary tipplers in England. 700 tipplers were built in 1939-40, but even with the capacity revised to 16 tons in line with the comparable coal wagons, carrying capacity was lower than desired. Improved capacity was provided by forty straight sided 23 ton wagons built by Hurst Nelson and Birmingham Carriage & Wagon in 1940. A 23 tonner, BSC 6114, is seen at Corby in 1976 only a couple of years before ore quarrying there ended. Unlike conventional private owner coal wagons which were requisitioned on the outbreak of war, and later nationalised, the tipplers were specialised wagons, so did not become part of BR stock, passing many years later from S&L to the nationalised British Steel.

Centre left: When invited to give evidence to the Ideal Stocks Committee, the British Iron & Steel Federation predicted a move away from the iron ore hopper towards the rotary tippler, and as the LMS had already shown the way, by sealing up the doors on a standard 16 ton steel mineral wagon, it was not surprising that the 16 tonner became the basis of a straight-sided tippler without side or end doors. Between 1951 and 1961, 9,590 Iron Ore Tipplers were built to five separate diagrams. Except for the final batch, all were unfitted, and until 1958, shared the 9 ft wheelbase of the 16 ton steel minerals. A pair of earlier vehicles, B385031 to Diagram 1/183, built at Shildon in 1955, and B382684 to Diagram 1/181, dating from 1954, are being loaded with road stone at Cliffe Hill Sidings on the Leicester to Burton line in April 1970.

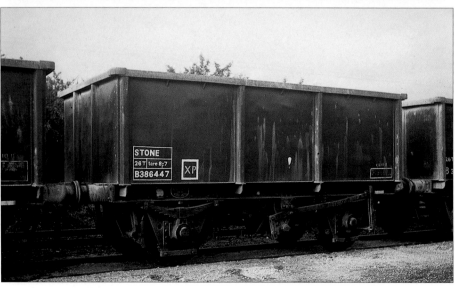

Bottom left: In 1958, a new tippler diagram appeared, for a wagon with a 10 ft wheelbase, but without automatic brakes. B386447, built at Derby to Lot 3091 in 1958, is seen in bauxite livery at Coltishall in Norfolk on 30th June 1972, after fitting with automatic brakes. The old IRON ORE TIPPLER branding has been replaced by a more prosaic STONE.

Opposite page bottom: A beautifully clean 47 179 eases a rake of loaded 23 and 27 tonners out of the exchange sidings at Corby in 1976. Owing to overloading problems with the early diagrams of 27 tonners, the body height was reduced by 6 inches on later deliveries, this variation is apparent between the fifth and sixth vehicles.

Right: For the final deliveries, built by Derby to Diagram 1/185 under Lot 3363 in 1960/61, the 10ft wheelbase, roller bearing axleboxes and 8 shoe vacuum brakes were specified. Stanier 8F No 48680, eases a rake of tipplers off the Stratford-upon-Avon & Midland Junction Line at Fenny Compton on 27th July 1964. In contrast to the black lettering patches on grey unfitted stock, the lettering is applied directly to the Bauxite paintwork of these Diagram 1/185 wagons. This rake of empties has come up from the hungry steel furnaces of South Wales, and will head south towards the Oxfordshire iron stone fields.

Right: By the early 1960s, the Stewarts & Lloyds tipplers were over twenty years old and more than 200 new wagons to BR Diagrams 1/183 or 1/185 were built. BSCO 25126, at Corby in 1976, was from a batch of 35 Diagram 1/183 wagons built by Metro-Cammell in 1962.

OPEN MERCHANDISE WAGONS

After coal wagons, one of the most numerous types of wagon in pre-nationalisation days was the 'open'. The term was all embracing, covering everything from a Container Flat with a vestigial body, through low sided Opens to the Highs, or 5-plank general merchandise opens. For ease in tracing the story of a distinctive group of wagons, Conflats and Lows have been covered in a separate section, and in this section we will only examine Highs and Mediums.

Above left: If one was to select a typical pre-grouping wagon, it would be a 5-plank drop door open and the oldest vehicle in this section is a London, Brighton & South Coast Railway designed 5-plank Open, which was one of 457 Diagram 1369 wagons transferred to the Isle of Wight by the Southern Railway. Although a Brighton design, construction continued under the Southern for a while. On the mainland, they would be a general merchandise wagon rather than a mineral wagon, but they served both purposes on the Isle of Wight. Most mainland examples had gone by the late 1940s, but about 150 were still in use on the island in the 1960s. Although the Cowes to Ryde line, had been closed for 8 months, S19072 was still in use at Newport on 29th October 1966, as periodic loco coal trips were still worked from Medina Wharf to Ryde.

Centre left: At the time that the Brighton wagon pictured above was being built, a number of companies had already switched to steel underframes and this process continued. By the 1940s, the differences between the 5-plank wooden bodied Highs produced by the Big Four were negligible. DM416335, at Bedford on 7th October 1971, was one of 3,775 Highs built to LMS Diagram D2094 between 1943 and 1946. Based on the standard 17ft 6in chassis, they differed from pre-war Highs in not having a curb rail, the ends of the floor sheeting being visible beneath the bottom plank on the sides. Originally unfitted, 416335 probably entered service unpainted, and later received BR light grey livery, traces of which could still be seen.

Left: Contemporary GWR Highs were similar. A Morton braked wagon, DW138961, in Departmental gulf red livery at Hinckley on 20th September 1971, was built to Diagram 0.37 in 1941, but has had the bottom two end planks replaced by channel section steel. This was a common modification introduced to combat the damage caused in rough shunting when steel or other heavy loads could break through the end of the wagon.

Above left: The LNER wagon shops were equipped to build and repair wooden underframes and it was not until 1938 that the company finally introduced a steel underframed version of the 5-plank High. E312938 is seen in gulf red livery at Northampton on 1st May 1976. As with the GWR High, Morton brakes are provided on this unfitted design.

Above right: Having resisted steel underframes until 1938, the LNER took less than 7 years to move to all steel wagons, with the celebrated High Steel of 1945. The earliest examples had wooden doors, but a steel door version soon appeared. One shortcoming with the early Highs was that it was impossible to knock nails into the body for securing ropes, so later batches had dimples in the body sides to accommodate internal rings to which ropes could be tied. LNER Steel High DE314624 was at Bedford on 7th October 1971. The unusual LNER vacuum brake gear shows to advantage in the low winter sun. Instead of the usual V

hanger located centrally below the door, there were two hangers on one side, and one on the opposite side, the vacuum brake cylinder being also on that side of the vehicle. Although the two V hangers on the near side are partially obscured by the door stops, the brake lever, V hangers and the linkage connecting them can be seen. This reversed the motion of the near-side brake lever in much the same way as the Morton clutch so that all eight brake shoes could be applied by the vacuum brake or by either lever. It was an efficient if complex system.

Below: The Ideal Stocks Committee expressed a preference for the LNER Steel High but accepted that it was unsuitable for certain traffic, including many foodstuffs. The Committee therefore advocated adoption of the Steel High, so far as was consistent with the wishes of traders. 2,000 Steel Highs were built to BR Diagram 1/037 in 1950, with a further

7,469 to Diagram 1/041 between 1950 and 1954. B483198, at Wellington (Shropshire) on 30th June 1976, was built as Shildon to Lot 2197 in 1951/52 and has LNER-style double V hanger brakes. Model railway experts sometimes tell us that loading a container in a High is unprototypical, but no one seems to have told this to the loader of BD49159B, a type BD container to Diagram 3/050, which was built at St Rollox in 1957!

The next wagon, 13 ton van B762829 is to Diagram 1/208 and was built under Lot 2595 at Wolverton in 1954. The third vehicle, B924270, was a short (32 ft over headstocks) Bogie Bolster E, built at Ashford in 1962 to Diagram 1/479. The Bogie Bolster E was a BR design owing little to pre-nationalisation practice and 1,200 were built in less than two years. Four bolsters were provided, the centre pair being fixed, but the outer bolsters could be placed in any of three alternative positions, slotting into the prominent sockets mounted on the outside of the bodywork.

Top: Although the Diagram 1/041 Steel Highs replicated the LNER design, BR developed the type further. The result was one of the most untidy opens ever built, the Diagram 1/047 and 1/049 BR Steels. Unlike the LNER design, the side sheeting stopped short of the planks which projected beneath the sides, whilst a bulbous reinforcing rib ran along the sheeting above floor level. The body was joined to the underframe by stanchions each side of the door and by the end stanchions, but this was inadequate for rough shunts, so ungainly end and side brackets were also required. B496765, photographed at Melton Mowbray on 30th July 1971, was built at Shildon in 1954 under Lot 2469, and is a Diagram 1/049 Steel Hybar, fitted with a sheet support rail, vacuum brakes and Morton hand lever.

Centre left: The LMS, GWR and Southern wooden bodied Highs were replicated as BR diagrams, the LMS design being the most numerous. B478587, at Hinckley on 20th September 1971, is a Diagram 1/039 LMS High Bar built under Lot 2179 at Derby in 1951. It has the corrugated steel ends of the late period LMS Open Merchandise wagons, LMS 8-shoe clasp brakes and Morton lever.

Bottom left; Rough shunting spelled trouble for the railways with traders submitting claims for breakages. To reduce these and promote better relations with their customers, the Big Four investigated shock absorbing wagons. The body was not secured directly to the underframe, but via longitudinal springs which allowed a little longitudinal motion. Shoc-Bar B721765, at Rugby on 28th June 1971, was built at Derby in 1951 under Lot 2180 to Diagram 1/040. It was one of no fewer than 11 separate diagrams for Open Shocks, and closely followed LMS practice. The longitudinal shock springs are hidden by the protective sheeting along the solebar. Shock wagons were identified by the three vertical white stripes.

Above: Before the war, the LMS was the only company to build the 13 ton Medium in large quantities. The most prolific design was Diagram D1927, of which over 10,000 were turned out between 1935 and 1949. The wagon on the right, DM472677, is a Diagram D1927 Open, which has been transferred to departmental duties in the tunnel maintenance train and is seen in gulf red livery at Rugby in June 1975. Beside it, DB457350, also in gulf red, is to the much rarer BR version, Diagram1/017, of which 397 examples were built at Wolverton in 1950. The similarity between the two designs is apparent from this view. After a few years in general traffic, most Mediums had been transferred to departmental duties by the mid 1960s. A couple of Diagram 1/017 Mediums survived in this capacity into the mid 1990s.

Above: The most prolific BR Medium was the Diagram 1/019 steel bodied Medium or Medfit. Between 1950 and 1955 Ashford works turned out 3,600 wagons to 6 lots. DB460239 was built under Lot 2351 in 1952 and transferred to the Isle of Wight in July 1986, along with sister vehicle DB461225, which had been built under Lot 2488 in 1956. Both are together at Sandown engineers yard in June 1994.

Bottom right: Bricks have long been conveyed by rail, but other than some 50 ton GNR/LNER bogie brick wagons, no specialist vehicles were built until 1950 when three experimental 4-wheel brick wagons were constructed. These had no effect on future development, as the need was for stock which could accept pallets of bricks loaded by a fork-lift truck. Surplus Medfits were modified with raised portions to the drop-sides and internal partitions, with end movement controlled by screw adjusters. Specialist Palbricks, with sides which could be lifted out, followed and more Medfit conversions took place to cope with a heavy but short lived traffic before brick manufacturers switched their business to the roads by the late 'sixties. B458049 began life as a Diagram 1/019 Medfit built to Lot 2235 at Ashford in 1950/51. It was modified to a 13 ton Palbrick A during the 'fifties and is seen at Marston Valley brick works on the Bletchley to Bedford line on 17th September 1971. It is branded PALWAG – a name which appeared on a number of wagons, but not in the wagon Diagram books!

The variable nature of the British climate led to a demand for covered goods wagons from the earliest days, but the low building cost of the open wagon and the simple waterproof sheet meant that many items which were better suited to vans were still conveyed in open wagons in 1948. BR built a high proportion of vans, compared to general opens to remedy this deficit, one Diagram, 1/208, running to over 19,000 examples by the time construction ended.

Top left: The Great North of Scotland Railway was the smallest of the Scottish pre-grouping companies and the only large community served by the railway was Aberdeen. Its main claim to fame was that uniquely amongst the significant railways of Great Britain, it never possessed a single 0-6-0 goods engine, relying instead on a fleet of 4-4-0s for freight as well as passenger duties. The lack of conventional goods engines was an eccentricity of the GNSR, but also a reflection on the sparse freight traffic north and west of Aberdeen. Predictably its total wagon stock of around 3,800 vehicles for 334 route miles was well below the national average, but they included some fascinating vehicles. Although considerably altered from GNSR days, this 4-wheel van, used to carry grain within Longmorn Glenlivet Distilleries Ltd, was recorded on 28th August 1978. As built, it was horizontally planked with external framing, but has received aluminium cladding, though the strapping and door hinges appear to be original. The steel underframe is unusual, as the channel section of the headstocks faces outwards, requiring wooden blocks between it and the buffer casings. The spring trunnions (the mounting brackets for the ends of the springs) are very deep, the handbrake is of an archaic type without a modern V hanger and works on a single wheel. Hooks for shunting by horse power have survived on the solebar just above the near axleguard. *Chris Nettleship.*

Centre left: One of the best known vans of all was the GWR Iron Mink, the first of which appeared in the days of William Dean, whilst a few examples survived into the post steam era on BR. Normally in plain GWR freight stock livery, a couple of Minks wore an eye-catching blue finish during the drab days of the Second World War. In the early months of the war, salvage drives were instituted, covering everything from aluminium pots and pans to garden railings and paper. They actually contributed little to the war effort, but went down in folklore. A Diagram V6 Iron Mink, 11152, of 1900, from the Didcot collection has been restored to record this episode and bears the correct number, 47305, of one of the two waste paper salvage vans operated by the GWR.

Opposite page top: The Great Western brown liveried stock, which included a range of vehicles for milk, fruit, fish, carriage trucks and scenery vans, occupied an ill defined niche between coaching stock and freight equipment. Some, such as the 50 foot bogie scenery van, or Monster clearly belonged in the coaching stock category. Others had more in common with wagons, and on some railways, would have belonged in the wagon category. This example GWR 2356, preserved at Didcot, is a Diagram Y2 Fruit B, built to Lot 638 in 1892. It displays the yellow lettering applied to the brown stock and the graceful GWR italic script for *Tons* and *Tare.*

Above: Another pre-nationalisation wagon popular with modellers is the GWR Refrigerator Meat van, or Mica B, on account of its dramatic white livery with red lettering. The GWR used telegraphic code words for wagons thus adding many strange names such as Mica, Toad, Mink or Beetle to the railway dictionary. 105860 is a Diagram X8 van, built in 1925 to Lot 921, and is preserved by the Great Western Society at Didcot, where it is seen in this 1989 view. Monochrome photographs suggest that the MICA B code and the running number are a little too fine, but in the absence of good pre-war colour views of Mica Bs, this merits inclusion.

Centre right: The Southern was much the smallest of the grouped companies and had a modest wagon fleet compared to the LMS or LNER, but it included many interesting vehicles. Perhaps the most improbable survivor is this former Isle of Wight Central Railway van No 87, which is thought to have been built at the IWCR works at Newport. It was to the unusual length of 18 feet, which was exceptionally long for the pre-grouping era. The wooden solebars are reinforced by external steel flitch plates. It was withdrawn in December 1925 and used by a farmer for over 60 years before being moved to Haven Street on the Isle of Wight Steam Railway in 1987. It was photographed in 1994, more than 70 years after the IWCR went out of existence, but traces of the IWCR lettering are visible to the left of the door, as is its original number 87, on the right of the vehicle.

Bottom right: The Isle of Wight became a haven for vintage wagons. DS46931, seen at Ryde St John's on 20th May 1964, was an LBSCR timber framed box van, one of three types sent to the island by the Southern in the 1920s and 1930s. Island stock had a habit of outlasting its mainland contemporaries by many years and a handful of LBSCR vans survived into the 1960s. DS46931 is in BR light grey unfitted stock livery, with black patches for the lettering. Painting styles differed between works and even painters. Although grey bodywork was standard on unfitted wagons, solebars could be black or grey. Steel solebars were usually black, but contemporary views of other IoW wagons show some wooden framed wagons with black solebars, and even black headstocks. The painting of the brake lever guard grey above solebar level and black below is unusual. This was usually black, regardless of solebar colour.

Left: The Midland Railway 16ft 6in van was developed from earlier 14ft 11in designs, and with its wooden solebars, outside framing and horizontal planking was typical of its era. This Diagram 363 Midland van ended its career grounded on the former Stratford-upon-Avon & Midland Junction section of the LMSR. I have included this view because it shows the framing of a Midland van to perfection, along with the ironwork. It also gives the modeller useful details of what a grounded van should look like. *Please* note that with certain types of van, particularly with wooden underframes, a grounded body includes the solebars and headstocks, without which it would fall to pieces. Lastly, the way in which tarred felt has been roughly nailed over the wooden panels, to keep the contents dry, would enable the modeller to hide a multitude of sins.

Centre left: The Midland adopted steel underframes as early as 1911 and, as MR influence was powerful within the LMS, steel underframes were standard for LMS vans too. From 1924 many thousands of vans were built with corrugated steel ends, and wood side sheeting. Early examples were vertically planked, but horizontally planked designs appeared from 1933. Diagram D2039 first appeared in 1940 and production continued until 1945. The majority were hand-braked only. M517768, seen at Lenwade, on the former Midland & Great Northern Joint on 13th July 1971, has been rebuilt with vacuum brakes and carries additional diagonal strengthening strapping on the body.

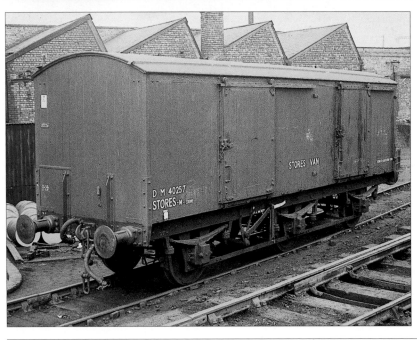

Bottom left: With our next vehicle, which began life as an LMS Express Fish van, we are entering the disputed territory between freight stock and non passenger coaching vehicles. Designated coaching stock when built, the surviving examples of LMS Diagram D2115 were redesignated wagon stock about 1965. DM40257 was built to LMS Lot 1445 at Wolverton in 1949 to a design which had first appeared three years earlier. They were an enlarged 6-wheel version of the early 1940s 10ft and 10ft 6in 4-wheelers. The body was an elongated version of the final 4-wheel design with an additional wagon-style sliding door. The chassis, though built to coaching stock specifications, was provided with a typical wagon brake lever. Three batches were built, one by the LMS and two by BR. All emerged from Wolverton works in crimson lake passenger livery with LMS style serif lettering, the 1949 stock being branded X-FISH to denote its express fish capability. Barred from passenger train workings in 1959, withdrawals commenced in 1964. DM40257 has been repainted in departmental olive drab, a colour introduced in 1963, for wagons used to carry locomotive stores from Crewe. It is seen here at Edge Hill shed in Liverpool on 10th February 1967.

Top right: The LNER was much slower than the LMS to adopt steel underframes or corrugated steel ends, but had done so by the late thirties. However, wartime shortages of steel and good timber forced the companies to improvise by using plywood sides and even ends for van bodies. E265484, at Moreton in Marsh on 10th September 1971, was a ventilated 12 ton plywood van built at Darlington in 1945. It is fitted with Morton brake lever, vacuum and hand brakes. A softwood board above the number panel was provided for labels.

Centre right: The development of the rail network permitted perishable foodstuffs to be transported long distances and fish, which had once been confined to coastal districts, became part of the diet of the city dweller. Fast trains conveyed the catch from the ports, and fish vans were amongst the earliest stock to be vacuum fitted. The high running speeds also prompted a switch from the traditional 9ft wheelbase to wagons with 10ft and later 12ft wheelbases. The final LNER Fish design, which did not appear until after nationalisation, had its wheelbase increased to 15 feet. The van was highly insulated with recessed sliding doors, and internal ice bunkers. As they were suitable for passenger train working, steam heating pipes were provided as well as vacuum brakes. Fish traffic would have declined anyway, but the rail strikes in the 1950s, which caused serious losses to the fishing industry, led to a collapse in fish traffic. Many fish vans were scrapped but some were transferred to general or engineering use. E75289, built at Faverdale wagon shops in 1949-50, was at North Walsham on 31st June 1972. It has lost its end ventilator and has been reclassified as, INSULATED VAN. It even bears the inscription, 'NOT TO BE LOADED WITH FISH'. The original Insulfish livery was white bodywork with a blue spot. Some conversions acquired the normal fitted-stock bauxite. As a modeller myself, I know that we usually want precise descriptions of liveries, but all I can say in this instance is that distressed white is probably the most accurate description which I can apply to the livery of this particular wagon!

Bottom right: The theoretical distinction between BR and pre-nationalisation stock is clear. W, S, M or E for the pre-nationalisation designs, and B for post-1948 designs. In reality that simple and logical definition falls apart. The LNER Insulfish just described was to a design which was prepared by the LNER, so has an E prefix, although none appeared until BR days. Our next wagon, B750083, seen at Fakenham on 11th July 1971, came from Wolverton and was also built in 1949 to BR Diagram 1/200. What is the problem? Diagram 1/200 was so like LMS Diagram D2108 that the uninitiated might think it was identical. The initiated knew it was the same. What makes the whole episode Gilbertian is that the 1,300 wagons had been ordered before nationalisation as LMS Lot 1456, to Diagram D2108. This was amended on 8th December 1948 to become the first ever BR wagon order, Lot No 2001, to Diagram 1/200. British Railways had merely put a new reference number on the LMS Diagram and issued it as their own.

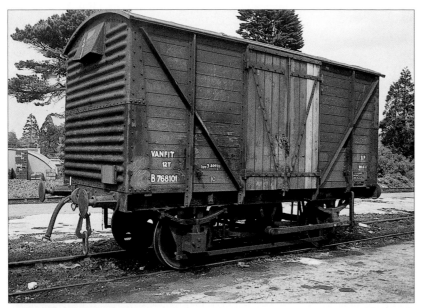

Left: The definitive BR van was that set out in Diagram 1/208, which was built continuously from 1951 to 1958, mostly at Wolverton and Faverdale, though there was a solitary batch from Charles Roberts. In theory to one design, the 19,000 wagons displayed many detail differences, with plywood or planked sides, ply or planked doors, Morton brakes or 8 shoe brake gear. B768101, which came from Wolverton under Lot 2707 in 1956, has planked sides and doors, the latter with diagonal strengthening, and is in pre-1963 livery, with unboxed load and number, but with tare weight in lower case and kgs at the left-hand end of the vehicle, and the 'Small print' subsidiary panel at the right-hand end where the tare would be in pre-1963 practice. Given the rusty ironwork, my opinion is that the dark bauxite colour is a well weathered form of the original paint applied at Wolverton in 1956, with lighter patch painting, unpainted wood on the door and a partial revision of the lettering. The van was photographed at Dorchester in May 1975.

Centre left: B774419, at Moreton-in-Marsh on 10th September 1971 was also built to Diagram 1/208 and came from Wolverton in 1957 as a part of Lot 2990. In this instance, we have a planked body with ply doors and no reinforcing strap. The many chalk marks on the body and the traders labels which have been applied to the van are typical of the period, and are commonly neglected by modellers. Sadly I must plead guilty here too! The decaying PARCELS and MAIL label on the right hand side suggests its last duties had been during the pre-Christmas parcels rush in 1970.

Bottom left: Given the diversity of Diagram 1/208, the decision to introduce a new diagram for vans with ply sides and doors seems strange, but Diagram 1/213 was the all-ply version of Diagram 1/208 and 3,699 vans were completed between 1952 and 1957. The first batch, commencing with B765481, emerged from Faverdale in 1952. B765307, seen here, should be a part of Lot 2367, (B764481-765480) which also came from Faverdale in 1952, but to Diagram 1/208, which was the all planked or planked/ply version. Although nominally to Diagram 1/208 this van is actually to Diagram 1/213. However, it gets worse, as the wagon plate stated the correct Lot, 2367, but gave Darlington 1956 as the place and date of its construction. The former NER Faverdale shops are of course at Darlington, but the discrepancy between dates and design are not easily explained. The next vehicle, B776565, was built at Ashford in 1957 under Lot 3023, it was thus part of the final order for Diagram 1/213 stock.

Top right: Diagram 1/224, with all ply doors and sides, was a development of Diagram 1/213, but with Oleo pneumatic buffers. Ordered as a part of the 1961 program, the 2,000 vans to Lot 3398 were the only wagons built to this diagram and the only conventional vans built for BR by Pressed Steel, whose previous contribution had been in the mineral wagon field. B786276, seen at Melton Mowbray on 30th July 1971, is actually plated 1962, but single year discrepancies were not uncommon on large orders such as this, where official records gave the planned introduction date of the first examples. A large programme could run on into the following year.

Centre right: As well as using Shock opens, BR built several batches of Shocvans. Apart from 50 built to an LNER design at Faverdale in 1948, and accorded BR Diagram 1/201, the first significant batch was 500 wagons built by Ashford in 1950 to Diagram 1/207 under Lot 2158. The design was prepared at Swindon as a plywood version of the GWR experimental Shocvans of 1936-40, with their distinctive twin end ventilators and sharply curved roof. Unlike the LMS Diagram 1/200 vans referred to earlier, the change in material did give them a recognisable difference from their GWR predecessors. B850145 is at Moreton-in-Marsh on 10th September 1971.

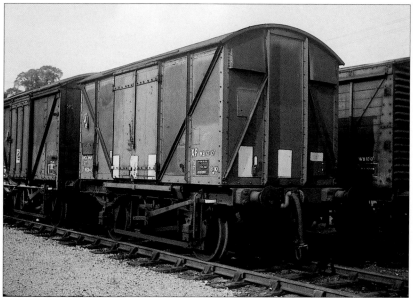

Bottom right: The introduction of pallets was one of the biggest advances in the handling of merchandise of all descriptions, slashing handling costs, breakage and pilferage. However it posed new problems for BR for the fork lift truck and pallet were too large to pass through the 5 foot doorways of most goods vans. The Palvan was introduced in 1953 to meet this new need. It had an 8ft 5in two piece door at one end of the vehicle, but in service was found to be unstable at speed. The next attempt to solve the problem was the Diagram 1/217 Vanwide, of which almost 2,000 were built to three lots in 1962. The doors opened outwards and then sideways to give a 9ft clear opening on a 17ft 6in body. Vanwide B783946 is seen at Toddington on 10th March 1999.

Above: This illustration highlights one of the problems facing the researcher. The Diagram 1/217 Vanwide was built to three lots, two from Wolverton and one from Derby. Lot 3392 of 894 vans (B783873-B784766), is customarily recorded as being a Derby batch, but B783946 carries a Wolverton plate. In several places in this book, we have noted minor differences between official records and the details recorded on wagon plates, but this is one of the more significant examples.

Below: In recent years, the model railway manufacturers have provided a variety of vans in private owner liveries, many of which have been fictitious, as the private owner van was a rarity. However, we shall look at a genuine example. In 1964, Standard Wagon built 96 long wheelbase pallet vans for Associated Portland Cement Manufacturers Limited. They were vacuum braked with a door which opened outwards and sideways to give a 9ft clear opening for fork lift truck access. With a number which might have been lifted from a James Bond movie, APCM BV 007 is at Gloucester in 1976. Under the TOPS scheme, the entire series were renumbered 6201-6296. They were used for bagged cement deliveries from Northfleet and Snodland, but with changing market requirements, were stored by 1982 and withdrawn a few years later.

Bottom: Now for the most colourful van in our selection, Transfesa Van 028 5 899-7, which was photographed at Ashburys in Manchester on 8th June 1977. Transfesa, or Transportes Ferroviarios Especiales SA, was formed on 2nd June 1943 to facilitate rail communications between the 5ft 6in (1676mm) broad gauge Spanish railways and the rest of Europe, by means of stock with interchangeable wheelsets. With the Anchor emblem prominently displayed on the side, this van was suitable for the Dunkerque-Dover train ferry service in pre-channel tunnel days and was provided with vacuum and air brakes with a change over valve, for operation in Britain or Europe. *Chris Nettleship*

CATTLE WAGONS

Although cattle traffic had fallen off appreciably between the wars, British Railways still required many thousands of cattle wagons in 1948 and were to build no fewer than 3,800 cattle wagons to four different diagrams between 1949 and 1953 when construction abruptly ended, as cattle traffic receipts were seen to be collapsing.

Top: The standard GWR 8 ton cattle wagon had emerged as early as 1888. With its channel section frame and angle-iron shell supporting the planking, it was an advanced vehicle for its day, and true-to-form, once the GWR had developed a sound formula, they saw no need for change. Save for the substitution of the original framed upper cupboard doors by a planked version extending to the same height as the safety rail, the design had hardly altered by 1948. BR built a large batch of wagons derived from LMS practice, a small batch based on a Southern design and then standardised on the GWR version, which became Diagram 1/352, but was upgraded from the 8 tons nominal capacity favoured by the GWR to the 12 tons popular with the LMS. B892264 was built to Diagram 1/352 under Lot 2126 at Swindon in 1950. Forty vans from Diagram

1/352 and the similar Diagram 1/353 were subsequently modified for use on the Dover-Dunkerque train ferry service. Requirements for conveyance by ship meant that the lower openings had to be blocked off whilst drains were fitted in the floor which led to a urine tank. These modified wagons became Diagram 354 regardless of their origins and B892264 was one of 26 Diagram 1/352 vans to be modified. When photographed at Rugby on 28th June 1971, it had been converted to an ALE VAN, a use which befell many cattle wagons in their declining years.

Above: B893394, built to Lot 2269 at Swindon in 1951, and B893969, built to Lot 2426 in 1952 are both at the cattle dock at Dingwall on the old Highland Railway Inverness to Thurso line on 9th July 1969.

Constructional differences between the two wagons include an additional row of drainage openings on the left hand vehicle. Unlike the previous wagon, they are 8 ton wagons to Diagram 353. One would expect considerable differences between an 8 ton and a 12 ton wagon, but can you spot any? The 11ft wheel-base, the length over headstocks of 18ft 6in, its overall width, door opening and many other dimensions were unaltered. The most significant change was a reduction in the overall height of just over 3in by means of a less sharply domed roof. A few other dimensions changed by up to an inch. The frame to the right of the door on the nearer wagon is the securing rack for the movable partition which adapted the wagon to short, medium or long configuration. This was to adjust the vehicle size to the number of cattle being carried.

In the dawn of the railway era, liquids were conveyed in ordinary open wagons loaded in casks or barrels The first railway tank wagon does not seem to have been built until 1865 when a D-shaped tank was mounted on top of a conventional wooden underframe. Very few tanks were built until the mid-1880s, by which time the D-shaped tank had been replaced by rectangular or cylindrical tanks mounted on wooden chassis. The Railway Clearing House issued its first specification for privately-owned rail tanks in 1887. These were periodically updated, the most important revisions coming in 1902, 1927, and 1939. Although hazardous substances for conveyance by rail are divided into many classes, two are of particular importance. Class A substances are highly inflammable liquids, such as petrol, with a flash point below 73 degrees fahrenheit. Class B are liquids, such as kerosene, diesel oil and light fuel oil with a flash point between 73 and 141 degrees fahrenheit. Heavy fuel oil, with a flash point above 141 degrees was commonly carried in wagons conforming to class B requirements.

Above: Ireland lacked both a large population and much heavy industry, thus tank wagons were rare. Because of the small number of vehicles needed and the unusual track gauge (5ft 3in), replacement was a low priority, and early examples survived into the era of colour photography. One of the largest UK operators, Shell-Mex and BP Ltd, had their own Irish subsidiary, whose class B tanker No 271, seen at the GNR(I) yard at Adelaide on the outskirts of Belfast on 24th August 1965, was built by the Midland Railway Carriage & Wagon Company of Birmingham. Not to be confused with the Midland Railway Company, this was one of several carriage and wagon builders located in the West Midlands. No 271 resembles a cradle design in which the tank rests not on transverse saddles, but on longitudinal wooden bearers, but careful study shows that the tie down straps, bracing rods for the end stanchions and crossed wire hold down ropes all pass between the tank body and the longitudinal, so the tank does not rest on the longitudinals. It is a saddle mounted tank with an additional L section angle iron bolted on above the channel section solebar for added strength. MRC&W built similar tanks for mainland use by Shell before the Great War.

Above: A close up view of the builder's plate on Smith & Forest No 1. Wagon builders fitted their own wagon plates, and private owner wagons additionally carried a registration plate which showed that they had been inspected and approved by a main line railway company prior to acceptance for travel on the rail network. In 1902, the Railway Clearing House rules laid down liveries for tanks. For class A liquids, which were highly inflammable, the tank was to be light stone (buff) with a horizontal bright red band running round the sides and ends of the tank. The solebars were to be bright red. Tankers for class B liquids were initially to be red oxide, and it is to the Oxide or Indian Red specification that the Great Western Society have restored this rectangular tank. Black was subsequently seen to be more suitable, as spillages of black liquids were less conspicuous. These rules remained in force until 1939 when owners were permitted to switch to aluminium tanks for class A liquids. During the Second World War, matt dark grey was adopted to make tankers less conspicuous to enemy aircraft, but aluminium was reinstated at the close of hostilities, giving way to light grey by the early 1960s.

Above: The Great Western Society have restored this rectangular tank wagon built by Charles Roberts & Co in 1898 to carry the livery of Smith & Forrest of Manchester. The company had been founded in the 1860s and until the First World War supplied road tar in tanks similar to 'No 1' Some of their tanks were painted indian red, whilst others were black, and the GWS have opted for the more colourful scheme. When built, this wagon would have had single side brakes, but has been modified to independent double side brakes at some time. Although long regarded as archaic, the last rectangular tanks were built shortly after the Second World War and a few remained in service until the 1960s. Apart from tar, they were also used for oil, so offer wide possibilities for the modeller for any period from the 1880s until early BR days.

Top right: Irish Shell and BP Ltd class A tanker No 2617, at Grosvenor Road goods depot, Belfast, on 23rd August 1966, is another Midland Railway Carriage & Wagon Co product, the starfish shaped maker's plate being carried on the solebars above the V hanger. The low evening sun picks out the brake gear and lower tankside details particularly well. Except for the crossed tie down wires, which were looped round the manhole mounting on the top of the tank, and which went out of fashion for new construction after 1927, this remained the standard design for rail tank wagons from before the First World War until 1944, when a new method of holding down the tank was approved. The central black inscription, NO LIGHT TO BE BROUGHT NEAR THIS TANK is the compulsory safety notice which appeared on all such wagons.

Right: Lancashire Tar Distillers No 507, a 14 ton cradle mounted tank was recorded at Partington in Lancashire on 14th September 1969. The positioning of the Commuted Charge sign on the end saddle, rather than on the left hand end of the barrel, is unusual. The slightly elevated vantage point shows the construction of the chassis and the short end footplates excellently. The wisdom of a black livery for black oil or tar products is also apparent, for the lettering on the left hand side of the tank is all but obliterated. It should read Lancashire Tar Distillers, 74 Corporation St, Manchester.
Chris Nettleship

Bottom: For many years, the RCH had demanded crossed hold down wires which were looped around the manhole casing, but in the 1920s decided these could be dispensed with. Some early tanks had been provided with longitudinal stays connecting the end stanchions. The demise of the crossed hold down wires in the 1920s saw a re-emergence of this feature, the stay being additionally secured to the tank at the mid-point to prevent sagging. This former Esso tanker, serving as a static oil store at Narborough on 6th July 1976, recalls this style and offers an unusual scenic possibility for modellers. It carries the CC sign and small yellow star at the lower left hand end of the tank and the large white star at the right hand end. The CC sign stood for the Commuted Shunting and Siding Rent Charge Scheme, whilst the small yellow star indicated the Commuted Empty Haulage Scheme. These agreements relieved owners of small but vexatious routine charges for shunting, siding standage and empty haulage in return for a lump sum payment. For the railways, it saved a vast amount of paperwork in calculating trivial sums, in return for a lump sum payment. The large white star was adopted by the RCH in 1913 for vehicles with oil axleboxes which could safely run at an average speed of 35 mph, unlike the coal wagon which, with its grease axlebox, was confined to slow moving freights.

Left: After rapid early progress, the tank wagon altered little from 1900 to the 1950s, by which time slow transit speeds were making rail less attractive to the oil companies. The first step in modernising the rail tank fleet came in 1957 with a joint project between British Railways, Esso Petroleum and Charles Roberts. The new wagons were provided with roller bearing axleboxes, vacuum brakes and a 15ft wheelbase for better high speed running. Esso Tanker 44291 is at Cosford on 1st July 1976 in class A (highly inflammable) colours. From the immaculate state of the wagon and the absence of a hazardous traffic label, it appears to be ex-works in what was known as a purged condition. Even when tanks were empty they were treated as hazardous, as petroleum vapour and air combine to produce an explosive mixture. A tank was only regarded as non-hazardous if it had been purged with inert gas to remove all traces of petroleum vapour. It has the grey tank with red solebars of a class A tanker and was one of the last 35 ton wagons built before the adoption by Esso of the larger monobloc tankers in 1963. When built, it carried the oval Esso symbol at the right hand end of the tank. These were removed in 1975-76 though its former position can be readily recognised by the securing lugs!

Above: The rapid demise of older stock meant that block trains of vacuum or air braked 4-wheelers and air braked bogie tanks became commonplace by the early 1970s.

The end of steam traction also saw an end to the requirement for barrier wagons between the engine and class A tanks. BR Standard Type 2, No 25 207 heads a rake of class A monobloc tankers through the down platform line at Rugby in June 1975. The 4 letter code reveals a class 6 working and that its destination is the Birmingham Division of the LM Region.

Above: Despite the advantages offered by the new 35 ton glw tankers, fuel traffic was still being lost to road and one of the few positive developments of the Beeching era was the realisation that new long term agreements were vital if the oil companies and BR were to invest with confidence. In 1963 most of the major oil companies signed the new agreements which led to the rapid build up of a fleet of monobloc 40-45 ton glw tankers.

To keep the length of rakes the same, so as to use the old terminal fillers, Esso also experimented with close coupled rakes. LMS designed class 12 shunter, 12 049, shunts a rake at Northampton on 6th December 1970. Because of the much fatter tank, the ladder had been displaced from its traditional mid-position, to one end, as can be seen on 6005 and 6038. The Esso depot was by Northampton No 4 box, some distance from Northampton Castle yard, so it is likely that one of these tanks has suffered some defect, causing the rake of six to be dropped off at the yard for attention by the Carriage & Wagon staff.

Above right: Apart from class A and B liquids, monobloc tankers were built for bitumen, for which they required heaters and for a new commodity, Liquid Petroleum Gas (LPG). The first experimental 36 ton LPG tankers were built for Esso in 1963, but LPG tanker construction on a large scale did not begin until 1966-67 with some 40 ton glw monobloc wagons by Metropolitan-Cammell Ltd for Shell-Mex and BP. The 40 ton glw tanker was designed in the vacuum brake era but the abrupt policy change in favour of air brakes for new construction created major operational problems and the need for wagons to be braked on one system and piped for the other to permit air or vacuum trains to operate. Shell-Mex and BP Nos 176 and 188 head a rake of 40 ton LPG tankers at Rowley Regis on 2nd July 1976. The sidings serving the Shell UK Limited plant beyond the wagons were once a major source of freight traffic on the former GW line from Smethwick to Stourbridge Junction, but were out of use by the end of the 1980s.

Above: Apart from petroleum products, many other chemicals and gases have been moved by rail. R Y Pickering built a series of pressurised liquid chlorine tanks for ICI between 1942 and 1954. ICIM 47484 dated from 1951, and after withdrawal in the 1970s, was presented to the National Railway Museum. Although anchor mounting was common in petrol tanks by the early 1950s, the older cradle construction, in which the tank rests in a longitudinal wooden cradle remained common for chemicals. When new, these wagons would have carried the class 2 hazardous chemicals colour scheme of a white tank with 6 inch red bands round the ends, but in later years, this was changed to a white tank with a broad orange horizontal band. *Mike Hodgson*

BULK COMMODITY WAGONS

Traditionally most traffic, other than petroleum, was carried in simple open wagons or vans, but in the last 50 years there has been a steady move towards specialist wagons for bulk commodities such as grain, cement, or fly ash.

Top right: The lightly populated and flat expanses of East Anglia are well suited to farming and from Great Eastern days, grain was an important annual traffic on the railways. The long GE branch from Wymondham Junction to Dereham and Wells was kept open as far as Fakenham for agricultural traffic for many years after the end of passenger services. Grain was handled at Ryburgh, just south of Fakenham. Five bulk grain wagons occupy the loop near the station. The nearest vehicle, B885327, is a BR Diagram 1/271 20 ton covered grain hopper, built to Lot 2925 by Pressed Steel in 1956. 570 wagons to this diagram were built between 1951 and 1959. The design closely followed that of the LMS grain hopper, the most obvious difference being the omission of glazed inspection ports in the ends of the BR version. The second wagon, W42320, is much rarer, for it is an ex-GWR Diagram V25 hopper of which just 12 were built, six in 1935, the remainder, including 42320 in 1936, to Lot 1259. An inspection door on one side provided access to the hopper for repairs. The appearance of such a rare GWR wagon in deepest Norfolk is indicative of how wagon stock became mixed under BR. For the modeller, it means that one unusual vehicle can be used to break up an otherwise uniform rake of stock.

Centre right: This portrait of the grain loading terminal at Wroxham, on the GE branch from Norwich to Cromer, taken on 17th June 1972, should be of considerable use to modellers, as it shows the loading pipes. The nearest wagon, B885364, is another Diagram 1/271 hopper from Pressed Steel. The second wagon is an ex-LNER timber bodied 22.5 ton bulk grain wagon, so once again we have the common variety and a much older wagon in the same scene.

Bottom right: Having studied the hopper in its natural environment, we will look at B885666 more closely. This was a Diagram 1/271 hopper to Lot 3234, built at Derby in 1959. The 50 wagons in this lot were the last traditional grain hoppers built by BR. They, and a previous batch of 50 wagons, were piped, unlike the earlier examples, and when photographed at Sandy on the GN main line in February 1976, theoretically carried bauxite livery, though the only evidence of this is the buffer casing and two small areas on the body. The remainder of the wagon is covered in rust and grime. This wagon, though in worse condition than the unpiped examples in the previous views, is indicative of the state of much of the BR wagon stock in the closing years of the steam era and the first 10 years thereafter.

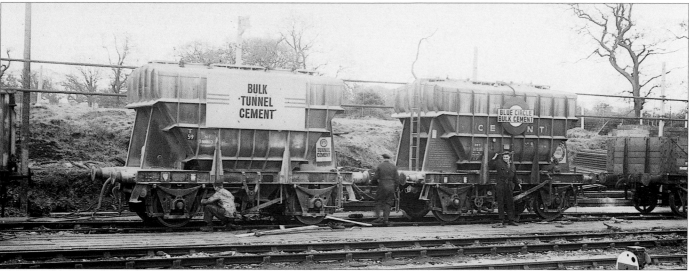

Top: Unlike grain hoppers which closely followed pre-nationalisation practice, designs for a bulk cement wagon had not even been considered by the Ideal Stocks Committee and it was not until 1954 that a prototype hopper cement wagon was ordered. Almost 2,000 were completed over the next few years, of which 30 were for salt traffic. They could be seen over much of the country. They were gravity loaded, but compressed air was used to help unload the dry cement powder, giving them their name of 'Presflo'. Blue liveried English Electric class 40, No 40 079 heads a rake of 10 Presflos near Kirkstall, between Leeds and Shipley, 20th May 1977.

Above: Two Diagram 1/272 Presflos are receiving attention in the wagon repair sidings at Brockenhurst on 24th April 1965. The wagon on the left, B888113, built under Lot 2863 at Shildon in 1956, is badged for Tunnel Cement, whilst the right hand vehicle, B873287, built in 1961 under Lot 3361 by the Gloucester Carriage & Wagon Company is plated for Blue Circle Cement. Compared to the earlier wagon, B873287 carries a lengthy inscription on the body end support and a different layout on the number panel. The older wagon is badged 20 tons and the later example 22 tons.

Right: The Presflo design was used in small numbers for salt and one Diagram 1/272 Presflo was converted for the conveyance of power station fly ash, becoming the sole example of a new Diagram 1/281. In its new guise, capacity was reduced to 17 tons. A production batch of Diagram 1/278 ASH 17VBs followed, as did three diagrams of enlarged 21 ton fly ash Presflos, Diagrams 279/280 and 282. The first 85 wagons were completed with vacuum brakes, but with the abrupt change to air braked designs, Diagram 280 and 282 wagons were air braked. They were unusual as they entered service with the B prefix of the old hand/vacuum fleet, although air braked, but later lost their prefix, bringing them into line with the other air brake stock. B873999, a Diagram 1/280 ASH 21AB of Lot 3531, squeals round the sharp curve from the Midland main line to the Melton and Peterborough line at Syston East Junction in September 1986.

LOWS AND FLATS

Machinery trucks or car flats and container flats are often classed with Open Merchandise Wagons, but this is misleading, as the load required chaining or securing in some way, rather than relying upon the wagon bodywork to keep it in place. I have grouped wagons with vestigial or no sides in this section.

Above right: From the dawn of railways, a demand existed for wagons to convey road vehicles and farm implements by rail. The well-to-do might wish to take their private carriage by rail, in which case it was conveyed by passenger train on a carriage truck, which was usually a long wheelbase vehicle notionally classed as passenger train stock. Less exalted equipment moved by freight train on machinery wagons. End loading docks were used for such freight, so the immediate recognition feature of a machinery truck, as they were often called, was a drop end. DS60568, at Ryde St John's Road in the Isle of Wight on 30th May 1964, is a former London Brighton & South Coast Railway Diagram 1661 machinery truck. First produced in 1892, the last examples were not built until the start of the Southern era in 1923. Unlike the majority of wagons built on a 16ft 6in chassis, the wheelbase is not 9ft, but 10ft 5in.

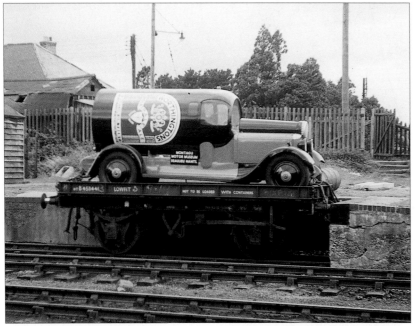

Centre left: BR inherited a mixture of machinery trucks for wheeled vehicles, some to passenger stock diagrams, other to freight stock diagrams. BR then added 3,150 of their own design within a few years. These were the Lowfits, the first 400 of which were wooden bodied, and the balance, to Diagram 1/002, steel bodied. B453441, one of the last ten examples of the final Lot 2998, was built at Shildon in 1957. On 26th April 1965, it was in the end loading dock at Brockenhurst station on the Bournemouth line. The load, a vintage Worthington Brewery van, belonged to the nearby Montagu Motor Museum at Beaulieu and provides a prototype for an out of period road vehicle on a wagon.

Bottom: By the early 1990s, less than 30 Lowfits survived, all in departmental use. Perhaps the strangest fate befell four lowfits which were rebuilt at Stewarts Lane in 1978 as adaptor wagons to move tube stock from the BR owned Waterloo & City line to Stewarts Lane depot for repairs. The conventional coupling and buffers at one end were removed, and a tube-style Ward coupler fitted well below headstock height to permit BR engines to move the tube trains. During the re-equipping of the surviving line on the Isle of Wight with former London Transport 1938 tube stock, two of the adaptor wagons, ADB452604 and ADB453241 were transferred to the island in 1992, where the latter is seen in August 1995. Although on the Island for three years, it retains its WATERLOO & CITY MATCH TRUCK branding. B453241 began life at Shildon in 1959 as a Lowfit built to Lot 2998 under Diagram 1/002. The rusty iron box on the decking is a tool cabinet which has seen better days as the lid is held on by rope.

Right: Containers are often thought of as a modern idea, but furniture containers existed 100 years ago. However, the container revolution really began shortly after the grouping when the LMS developed a variety of designs for different types of goods such as furniture, bicycles, general merchandise, refrigerated foodstuffs and even minerals. This 1930s LMS publicity card, promoting the household removals service, depicts K1, the original LMS Furniture Removal Service container of 1932 in crimson lake livery. Production type K containers, but not the prototype K1, carried the slogan 'Household Removals and New Furniture Traffic' below the 'Estimates Free' lettering, but otherwise were similar.

Centre Right: As the Big Four had made great efforts to promote container traffic before the war, it is astonishing that the Ideal Stocks Committee did not recommend the construction of container wagons, or propose a standard design. However, Swindon and Wolverton were already producing the first 400 Conflat-As and within a decade, there were 20,000 of them. The design was based on the GWR Conflat which, with its chain pockets on the solebar, could easily be mistaken for the BR version. B737759 is to Diagram 1/069 and was built to Lot 3107 at Ashford in 1958. The rated capacity of this diagram varied between 11 and 13 tons. The container, BK9005B, was a Diagram 3/127 BK container built to Lot 2830 at Earlestown in 1956.

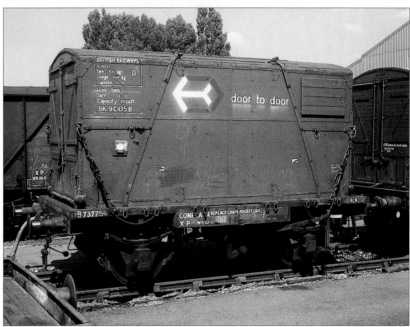

Below: The short wheelbase of the 204hp class 03 and 04 shunters meant that they could disappear from a track-circuited signal box diagram when negotiating pointwork. A number operated with a runner to avoid this problem. 03 081, previously numbered D2081, was photographed at King's Lynn in 1977 when running with (B)708135, a former Conflat A, built at Ashford in 1957 to Diagram 1/069. The yellow departmental livery shows the solebar details to perfection.

Above: Compared to the 20,000 Conflat As, the next most prolific design was the Conflat L with around 2,000 examples. They were designed to carry three 4 ton limestone containers. Early examples resembled a Lowfit with vestigial drop sides and ends and a long chain pocket, but the most numerous type, the 1369 wagons to Diagram 1/068, consisted of an underframe, floor, corner brackets to hold the containers in position, chains and chain pockets. B734331, built under Lot 2973 at Swindon in 1957, was photographed at *Swinden* lime works on the Grassington branch near Skipton on 28th June 1970. The nearest container, L71351B, was built at Earlestown in 1961, whilst the other two, L70061 and L70145, both came from Earlestown in 1958.

Centre left: Match trucks were flat wagons which were used where loads overhung the ends of the wagons on which they were conveyed. BR-built match trucks comprised just an underframe and decking, but the GWR provided a vestigial side, so that the vehicle resembled a conflat, and unlike the BR version, it was capable of conveying containers. In 1950, BR started a separate Internal User wagon fleet, with each region having a separate number series. It was intended for stock which was confined to a particular depot and which would not work on running lines. This wagon, 082007 in the Southern Region 08xxxx series, was rebuilt at Derby in 1961 from a GWR match truck W115493, for use at the wagon repair sidings at Brockenhurst, where it is seen on 18th July 1972.

Above: The normal wagon has 4, 6 or more wheels, and earns its keep by moving its load around the country. Our next wagon can only be used when it has but two wheels and is not merely incapable of movement, but any movement would be catastrophic. Ministry of Defence (Army) No 95004, was one of five Ramp wagons built for the War Department by the Southern Railway at Ashford in 1940 and 1941, and was a 17ft 6in long flat wagon with detachable axles. It was used to unload tanks or other military equipment at locations where no end loading facilities existed. When in use, one end would be jacked up, the wheels removed and the buffers swung out of the way. A rake of flats or wells would then be shunted up to it, and vehicles manoeuvered over the short ramp.

Coded PXP, the miscellaneous section under TOPS, MODA 95004 is seen at the Central Vehicle Depot at Ludgershall. My father first got to know this area as a newly commissioned Second Lieutenant stationed at Tidworth in 1938 and took this view during a nostalgic visit to the base in July 1991, less then 3 months before he passed away. As a past Medical Officer to the 1st Royal Tank Regiment, he was deeply appreciative of the kind reception accorded to him at the depot.

RAIL AND TIMBER WAGONS

This classification stemmed from railway statistical returns and covered a diverse group of stock, including 4-wheel and bogie bolsters, steel plate, strip and coil wagons and rail carrying stock.

Above right: The earliest wagons were short wheelbase 4-wheelers, but the need to convey long loads, such as tree trunks or rail, soon arose. The answer was to carry the load on movable beams or bolsters on two or more short wagons. By the 1930s, the benefits of long wheelbase double bolster wagons, or bogie stock, were apparent, but even BR built a number of single bolsters and in the Isle of Wight, a batch of former LBSCR single bolsters survived to the end of steam operations. 59050 built in 1910 to LBSCR Diagram 1616, was sent to the Island in 1930 and sold to the Wight Locomotive Society in 1967. It was restored to SR livery in 1985 and is depicted at Wootton in 1994. A view of two sister wagons, 59043 and 59044, both in different BR liveries, appears in my book *Rails in the Isle of Wight - A Colour Celebration*, also available from Midland Publishing.

Right centre: The Midland Railway Diagram 827 bogie bolster was a much more impressive vehicle. It was 45 feet over headstocks, with a steel underframe and substantial trussing and was rated to carry 25 to 30 tons. Production continued under the LMS until 1934. DM117519, at Longsight on 20th March 1976, is a Midland example from Lot 946 of 1920 and is still fitted with the MR brake wheel at the far end. Despite their size, such stock had to be able to negotiate sharp curves, and Diagram 827 bogie bolsters could be worked around a 1 chain radius curve (1 chain measures 22 yards or approximately 20 metres). Such curves were only to be found in a few dockyards, goods depots or private sidings, for a curve of under 10 chains radius on a running line had to be check-railed. In modelling terms, 1 chain radius equals approx 18 inches in 0 Gauge and just over 10 inches in 00. *Chris Nettleship*

Below: The LNER Quint D, so called because of its five bolsters, first appeared in 1929. Identical wagons, of 40 tons capacity and 52ft over headstocks later appeared on the LMS. E306462 was a Diagram 205 bolster, built by the Tees Side Bridge & Engineering Works of Middlesbrough in 1948 and is seen at Hinckley on 20th September 1971.

Considered to be 'Specially Constructed Vehicles' until the advent of BR, they were transferred to the general diagram book after 1948. The reporting name Bogie Bolster D was adopted for the 40 tonners inherited from the LMS and LNER and also for BR built examples. The 'CL' symbol marks the centreline of the vehicle.

Above: BR Diagram 1/472 was a BR re-issue of LNER Diagram 205, and the first batch of 200 wagons, B941150-941349, built by Tees Side Shipbuilding & Engineering in 1950, even had LNER diamond framed bogies as well, though later batches had BR/GWR plate bogies. B941282 came to grief in Rugby, where it is seen clear of the tracks in July 1972.

With the body dumped at an angle, I found this elevated viewpoint, which shows what the decking looked like, and how it had weathered, very useful in building some Bogie Bolster Ds for our own layout. The 16 tonner in the background, B116998, is to the usual Diagram 1/108 and came from Head Wrightson in 1952.

Above: This line up of stock loaded with girders from a dismantled bridge shows the largest of the rail and timber wagons, the Borail, or Bogie Rail wagon, in use at Leicester on 6th July 1976. The girders may well be from the 'Ferodo Bridge' over London Road in Coventry, *en route* to Vic Berry's scrap yard in Leicester. The nearest vehicle is an LMS long wheelbase Brake to Diagram D2036 or D2068. The main difference was that D2068 vans were provided with a longitudinal weight box which extended below the solebar. The high angle of the sun means one cannot be sure if there is a weight box or merely deep shadow, though my impression is of a box. The next vehicle is a

Conflat A in use as a match truck, as the bridge girders overhang the ends of the carrying wagon. BR had built a considerable number of dedicated match trucks, but most had been scrapped by the early 1970s, and a Conflat A, which had other uses, was an excellent substitute. The first pair of girders are carried on a 50 ton fish-bellied Diagram 1/483 Borail MB or EB, built in 1960/61. The second pair of girders are on a Diagram 1/481 Borail WG of 1959, with the older trussed construction frame. The final vehicle is another Conflat A. These two vehicles display the different design philosophies between the shallow framed but heavily trussed Borail WG and the deep

fishbelly girder of the Borail MB/EB. The Borails, although classed as rail and timber vehicles, would have fitted more logically into Specially Constructed Vehicles on account of their size and the regional wagon control provisions which applied to them. Whilst many wagon types had a single letter suffix, such as Conflat A or a Palbrick C, the two letter code indicated the region to which specially constructed wagons were allocated for control purposes, the regional headquarters keeping track of the whereabouts of such wagons. The vehicle in the sidings to the left is a Diagram 1/575 Mermaid or 14 ton side tipping ballast wagon.

Above: Perhaps the most remarkable tribute to the design of the Bogie Bolster D, was that between 1975 and 1979, no fewer than 1,251 of the unfitted BR Bogie Bolster Ds were rebuilt. All but three received modern Y25C bogies and air brakes. Their carrying capacity was revised from 42 to no less than 57 tons. With an 80 ton gross laden weight, No 950821 was rebuilt to Diagram DB006D, as a part of Lot 3965, at Shildon in 1978 and is depicted in April 1998 at Wolverhampton Steel terminal. The old chain loops on the solebars have gone, along with the securing chains which held the load in place. Today, polyester straps, tightened by small ratchet tensioners, do the same job more efficiently. Six bolsters instead of five are now fitted to each wagon, but otherwise we are looking at a BR design of the 1940s, itself a reissue of a pre-nationalisation diagram, which is still giving good service half a century later.

Below: The LMS had built Long Low wagons for conveyance of sheet steel and similar loads, as had the LNER, the latter calling them Plate wagons. BR adopted the LNER name and built wagons to both LMS and LNER designs, before producing a vacuum fitted version of its own. B931595 was built at Shildon in 1952 under Lot 2327 to Diagram 1/431, which was the BR version of LNER Diagram 123. Starting life as an unfitted plate wagon, by the time it was photographed at Warsop in 1976, it had received air and vacuum pipes and had been transferred to departmental stock for use as a reach wagon at Warsop. Part of the side has been patch painted in the short lived Engineer's red livery of the early 1960s to carry its revised number, function and allocation. The narrow wooden battens, which held the plates clear of the floor to permit chains or slings to be passed under them, can be seen.

Above: B934745 is one of 1,500 Diagram 1/434 Plates, with vacuum brakes and clasp brake gear, built at Shildon under Lot 3223 in 1959/60. It is under load, with steel billets rather than plate, at Derby on 6th July 1976. It has received Oleo pneumatic buffers and has been re-coded Plate VB. The livery should be bauxite, but is actually rust and grime.

Right: Tube traffic was important even before the war and the LMS, LNER and GWR all had their own designs for this traffic, most of which were perpetuated after 1948 until BR types emerged. British Steel Corporation D33, a GEC Traction diesel hydraulic of 1974, shunts a rake of Tubes in the BR exchange sidings at Corby. The leading wagon, B733344 is to Diagram 1/448, which, at 32ft over headstocks and with an 18ft 6in wheelbase, was an elongated version of the LMS design. It was built to Lot 3332 at Derby in 1961.

SPECIALLY CONSTRUCTED VEHICLES

Specially Constructed Vehicles arose out of the railway's original legal status as a common carrier, under which they were obliged to move any load offered to them if it was physically possible to do so. For this reason, they needed vehicles capable of taking heavy slabs of metal, large transformers, gun barrels and other awkward, heavy or odd shaped items. Under BR auspices Specially Constructed Vehicles had their own diagram book, and long before the days of TOPS, the Total Operating Processing System introduced on BR in 1972, such vehicles were specially reported to district control offices.

Above: The Low Machine wagon or Lowmac, as it was christened by BR, was the most numerous of the special wagons. Its primary purpose, to carry large road vehicles, farm implements or machinery, is perfectly recaptured in this scene at Ashchurch on 1st July 1976. Now branded as a Lowmac EQ and carrying an Army Bedford truck between bases near Wool and Ashchurch, E230917 started life as an LNER Mac NV to Diagram 143 in 1939. Originally rated to carry 20 tons, the Mac NV was redesignated to convey 22 tons during the war, but this vehicle has been reduced to 20.5 tons.

Below: The build up for D-Day prompted large scale construction of wagons capable of taking the heaviest Cruiser or Infantry tanks of the day. As some of these, such as the American designed Sherman tank were too tall to clear the loading gauge if mounted on an ordinary flat, this necessitated a heavily built 50 ton bogie well wagon. They were ordered by the Ministry of Supply and at first bore an MS prefix to their numbers, which ran from 1 upwards. They were later transferred to the Ministry of Defence and given a new number series and latterly a TOPS number. MODA 83082 was built by the Gloucester Carriage & Wagon Company and is at Llandovery on 29th June 1974. Although customarily associated with the movement of tanks, they were used for any large piece of equipment such as with this MOD roller No 720374.

Right: The wartime need for heavier loads led to an improved Lowmac with deeper side members. This was the 25 ton Mac PV, reclassified by BR as Lowmac EP, the first letter in the suffix indicating that its movements were to be reported to Eastern Region wagon control. After the war, BR adopted the same design for their own Lowmac EP, which became Diagram 2/242. B904503 was one of 38 wagons built to Lot 2187 by P W McLellan in 1950. This study, at Gloucester in July 1976, shows the securing chains and timber chocks which were placed each side of the tyres of road vehicles.

Above: The ambition of the Kaiser to wrest control of the sea from Britannia in the early 1900s led to a massive arms race. As the Royal Navy built ever more powerful battleships to counter the new German fleet, there was a need for rail wagons to convey the massive guns, which increased from 11 to 13.5 and then 15 inches. In 1909, the GWR built a set of four close-coupled 6-wheel wagons to Diagram A6 to convey such loads. Each individual vehicle could carry 30 tons and they were given the reporting code Pollen E. They were numbered 84997 to 85000. In 1930, they were divided into two twin sets, but could be brought together for exceptional loads. Chris Nettleship saw the set at York on 9th April 1977 and photographed DW84997 and DW85000. By this time they were in departmental stock and were to be found at South View Yard, York. *Chris Nettleship*

Centre right: Coded Girdwag by BR, it was not until the 1990s that the four vehicles were finally retired, giving them a service life of almost *ninety* years. Happily all four have gone to the Great Western Society's centre at Didcot, where I took this elevated view of 84997 in April 1998. The similarities with locomotive tender frames are apparent, as is the Churchward brake, the box-like wheel splashers and the bearing plate for the load.

Right: Wagon books invariably concentrate on side views, pictures showing details of the decking are rare. I hope modellers will find this view of the decking and swivel plate on 85000 of help. Although the wagons have been recently refurbished, there is a good deal of rust to be seen, the immaculate wagons which all too often grace model railways are far removed from reality.

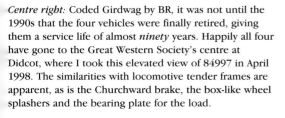

Left: To most people a trolley is a small item of furniture or something found in supermarkets, but to the railwayman, it was a highly specialised wagon, some of which were very large indeed. The Flat Trolley, or 'Flatrol', as the LNER and BR called them, had a depressed centre section, the load being carried on the flat floor itself. Unlike the Lowmac, with its sloping ramps for vehicles to drive on and off, loads were craned on to a Flatrol. DB900408, at Longsight on 14th June 1976, is a 40 ton Diagram 2/511 Flatrol MHH, built at Derby in 1949 to Lot 2005. The second vehicle, DM117519, is a Midland 30 ton bogie bolster, illustrated in more detail on page 63 of this book. *Chris Nettleship*

Left: The Weltrol was similar, but the load was carried on top of the side girders, or on transverse girders, and the floor, if present, was not of any structural importance. 20 ton Weltrol MC B900805 was one of ten wagons built to Diagram 2/730 at Derby in 1949. Four similar wagons, to Diagram 2/746, were constructed in 1959-60. The load on this occasion is a rather special one, the Welshpool & Llanfair Railway's locomotive, GWR No 822, *The Earl*, seen at Rainhill station on 24th May 1980. *Chris Nettleship*

Left: With 25 ton axle loads permitted over much of the modern rail network and 102 ton glw bogie mineral wagons or tankers commonplace, it is easy to forget that until the 1970s, the normal wagon was a 4-wheeler with 10 tons carrying capacity and a tare of 6 to 8 tons. Even today the 120 ton Transformer MB of the LMS or the 135 ton Transformer MC of BR would count as large wagons, in their day they were giants. The last and largest were the two Transformer MCs, B901800/01 of Diagram 2/470 built by Head Wrightson in 1952. They were 89 feet in length and the ends of the load beams were carried on cradles mounted on shorter beams which connected the two six-wheel bogies at each end of the wagon. The main girders could be moved inwards or outwards, or even shifted off centre to clear bridges. Movement of a load entailed a vast amount of preparatory planning, and might require the removal of some structures such as signals or lineside equipment. One of the Transformer MCs was recorded on 8th June 1977 at Ashburys, Manchester, where it was kept to be convenient for the Metrovick works at Old Trafford. *Chris Nettleship*

SPECIALLY CONSTRUCTED VEHICLES

DEPARTMENTAL STOCK

The Civil and Mechanical engineers had an almost insatiable appetite for elderly traffic wagons, sometimes with bizarre modifications. This mix of ancient wagons adapted from traffic duties, and purpose built modern stock, often constructed in small batches, made for great diversity. This variety, plus the esoteric reporting codes of an aquatic nature, such as Salmon, Dogfish, Mermaid or Shark means that departmental stock is always popular with modellers.

Top right: Until 1936 the LMS had sufficient pre-grouping sleeper and ballast wagons not to bother with any standard designs, but by nationalisation 459 sleeper wagons had been built along with 3,160 drop-side Ballast Opens. In each case, there were two diagrams, a more robust pre-war design and a wartime version with thinner planks. DM741374, at Rugby in August 1969, is a Diagram D1954 ballast wagon with two drop side doors, a centre pillar on each side, and drop ends. Livery is hybrid, with part of the vehicle in 1948-1960 departmental black and the rest in 1960-62 gulf red. This short-lived colour scheme was sufficiently similar to fitted stock bauxite to have confused railwaymen, who wondered why an unfitted vehicle should be in fitted colours. It has given similar headaches to modellers.

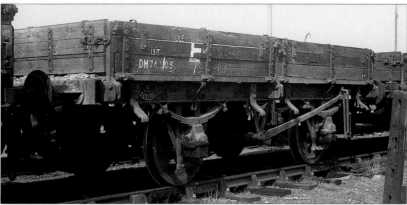

Below: The sleeper wagon was very similar to the ballast opens, but had fixed ends. DM749267, depicted at Windermere on 19th September 1971, was a Diagram 2098 thin plank Sleeper wagon, built at Derby in 1947 It is in gulf red livery and is branded HADDOCK.

Above: The wartime version of the D1954 ballast wagons, built to Diagram D2095, is represented by DM743105, seen at Marston Valley brick works on the Bedford line on 17th September 1971. Built at Wolverton to Lot 1479 in 1947, the poor condition and warped state of the low grade wood is apparent, as is the hybrid livery with gulf red, black and unpainted wood. The large E to the left of the number is a relic from the LMS Engineers livery, and may be the original lettering applied in 1947. Under BR the 3-plank drop side/end ballast wagon was code named Sole.

Left: The Butterley company built a single batch of 50 Haddock 12 ton sleeper wagons for BR in 1950. They were to the LMS design, but given BR Diagram 1/621. DB995040 is at Windermere on 19th September 1971. Except for the end plank in gulf red, the rest of the woodwork is in 1948-1960 black, but the lettering is boxed in post 1963 style, correct even down to the lower case tare lettering. By this time, the official colour for Engineers Department repaints was olive drab. Unlike traffic wagons where home station branding was rare, many ED wagons were branded for a particular depot, evidenced by the 'RETURN EMPTY TO DITTON SLEEPER DEPOT' plate carried by this wagon

Centre Left: Between 1896 and 1945, the GWR built over 2,500 drop door or drop side all steel ballast wagons of various sizes. DW80725, at Honeybourne on 10th March 1971, is a 10 ton drop side ballast wagon to Diagram P15, and is in BR departmental black livery. Note the brown road grime on the running gear and stone dust staining on the curb rail and on top of the buffer casing where stone dust settles during loading. The buffer faces are rusted, but black with grease on the centre, a seldom modelled feature! A similar vehicle, 80668, is preserved by the Great Western Society at Didcot in GWR engineer's black livery.

Bottom: BR copied the small GWR 10 ton drop side open, applying the codename Starfish, and also built 14 ton opens called Tunny, but the definitive dropside ballast wagon was the Grampus of which almost 5,000 were built between 1951 and 1961. The two top planks from the ends could be removed and stowed in racks below the underframe, whilst the lower portions were hinged. DB986284, at Rugby on 28th June 1971, was built by the Gloucester Carriage & Wagon Co to Lot 2885, Diagram1/572, in 1954, and is on a through working from the Southern Region. The adjacent vehicle, B483741, is a Diagram 1/033 drop side High built at Ashford in 1949 to a most peculiar Southern design. Full length dropside Highs have always been rare, due to the size and weight of the drop side and many of this batch were quickly transferred to departmental duties, as in this case, with a branding 'ON LOAN TO CIVIL ENGINEER SOUTHERN REGION'.

Right: The Isle of Wight was one of the few areas where dropside Highs found favour, and a number of Diagram 1/033 Highs were transferred to the island in 1970. Later replaced by conventional hoppers, four Highs survive in the Isle of Wight Steam Railway's collection. B483700 received a major refurbishment in 1993 and was seen at Wootton the following year. The fictitious Hocknulls Coal & Coke Merchants Private Owner livery is in recognition of the work being sponsored by a local coal merchant. The purist may complain, but at least the wagon has survived, enabling us to study these strange traffic wagons which spent most of their lives on engineering duties. It also provides an opportunity to remind modellers that not all wagons on preserved lines carry authentic liveries, but there is usually a good reason for this.

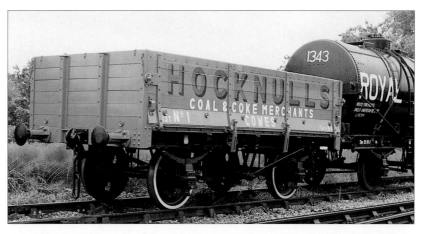

Above: In 1928, the LMS acquired a few 25 ton ballast hoppers from the Leeds Forge Company to Diagram D1800 with end platforms for operating the ballast door control wheels whilst the wagons were in motion. Four years later Metropolitan Cammell supplied some 25 ton hoppers with side wheels and no end platforms to Diagram D1804. These two diagrams became the basis of two BR designs, both of which received the designation, Trout. Less common than the later Catfish or Dogfish, they were the largest BR 4-wheel ballast hoppers.

Total production was under 200 wagons. BR Diagram 1/580, which was a copy of the LMS D1800 Leeds Forge wagons, was for 113 vehicles with an 18ft 0½in hopper, three discharge doors and an operating platform at one end. Diagram 1/581 followed the D1804 Metro-Cam wagons of 1932. The hopper extended the full length of the underframe, leaving no space for an end platform, but with the greater length, the hopper was over a foot lower than on Diagram 1/580! The wheelbase was reduced by 6in and instead of left, right,

and centre doors, two doors were provided on each side worked by hand wheels below the solebar. Two lots were built in 1949-50. DB992020, in the sidings at Nuneaton Abbey Street on 30th September 1971, was one of 24 Trouts to Diagram 1/581 built by Tees Side to Lot 2027 in 1949. This highlights the absurdity of two such different designs with the same code, when quite trivial differences had different codes. Eventually BR recognised the paradox too and the Diagram 1/581 Trout was redesignated, Gannet.

Left: The Herrings were another instance of different designs with the same code. The Diagram 1/582 Herring was a 20 ton design with a full length hopper based on GWR Diagram P22. It had side operating mechanisms. The Diagram 1/584 Herring, of which 100 were built by Metro-Cammell, had an end operating platform and the usual off centre hopper. In common with other early 1950s Metro-Cammell Mackerels and Trouts, it featured reversed channel section solebars, with the channel inside, and reversed side stanchions as well. DB992444 was one of three examples still extant on the Isle of Wight, though out of use, by 1994. Stored off the rails in the Civil Engineer's yard at Sandown, this is the perfect prototype for the modeller who has too many wagons to put on his layout!

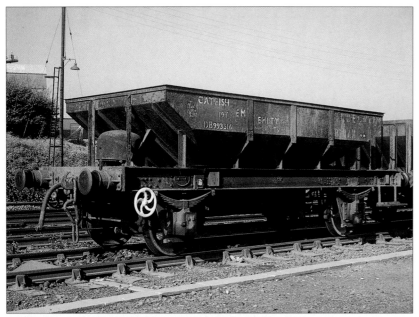

Left: Catfish DB993516, recorded at Nuneaton Abbey Street on 30th September 1971, offers a useful comparison with Trout DB992020, seen on the previous page. The 19 ton Catfish, to Diagram 1/586, was built by Metro-Cammell in 1959 to Lot 2929. Although the wagon is 2ft longer over headstocks, the hopper is almost 3ft shorter than on the Trout. The vacuum brake cylinder is mounted above the near end platform, whilst the control wheels for the hoppers are worked from the opposite end of the vehicle. Both wagons are in weathered 1948-1960 departmental black livery.

Centre left: In 1959, an enlarged version of the Catfish appeared. The Dogfish, Diagram 1/587, shared the same wheelbase, length over headstocks and over the hopper, as the Catfish, but the hopper was over one foot deeper, raising the capacity from 19 to 24 tons. DB992838, at Syston East Junction in September 1986, was one of the initial batch of Dogfish built by Charles Roberts to Lot 2819 in 1959. It carries TOPS code ZFV, and has had its cubic capacity raised from 527 to 634 cu ft by 9 inch deep additions to the sides. This was for the conveyance of slag ballast which is lighter than stone.

Below:. Although LMS influence was substantial, BR adopted practices from all of the Big Four. The GWR had built ballast hoppers in Victorian times, but at the start of the 1930s acquired 14 ton side-tipping ballast wagons to a design by the Metropolitan-Cammell Carriage & Wagon Co. The design re-appeared under BR, with 89 hand braked wagons built under three separate lots by Metro-Cam in 1952-56 and a further 200 fitted examples in 1959-61. DB989483 and DB989516, at Nuneaton (Trent Valley) in November 1975, are both fitted Mermaids to Diagram 1/575, built under Lot 3330 in 1960. Apart from the design, which copied the GWR stock, the 'CO' in a circle is a reminder of the Great Western branding of wagons used on construction work.

Above: The LNER developed 40 ton low-floor 65 foot wagons with multiple section drop sides for rail, sleepers, and ballast. Under BR, they were coded Dolphin, and were joined by BR built examples. A 50 ton version, the Sturgeon, to Diagram 1/638, and the Sturgeon A, to Diagrams 1/645 and 1/647, followed. There were seven drop sections on each side and with 2ft 6in wheels an exceptionally low floor height was feasible. DB994871, built to Diagram 1/645 under Lot 2614 by Head Wrightson, is at Windermere on 19th September 1971. Officially recorded as a 1954 wagon, the maker's plate gave a completion date of 1955. Such discrepancies were not unusual with freight stock where completion of a batch might be spread over many months.

Centre right. The advent of continuous welded rail necessitated suitable stock to convey it. The standard LMS 17ft 6in wheelbase chassis found a fresh lease of life on the early continuous welded rail trains. This channel-section chassis had been used on a variety of different wagons including Double Bolsters, Plates and Tube wagons. DM288205 was probably an LMS Diagram D1674 Double Bolster whilst DM726742 was a Diagram D2105 Double Bolster. They have lost their side rails and original bolsters but have gained new square section steel bolsters at the midpoint, with stanchions secured to the outside of the chassis. The rake is on the relief line at Hereford on 30th June 1976

Below right. One of the best known permanent way wagons is the Shark ballast plough brake van. This served as a guard's brake van on ballast trains, but when the plough attachment was lowered, spread the ballast which had been discharged from the hoppers a few moments previously. BR acquired two small batches of ploughs to former LMS and LNER designs before developing its own Diagram 1/597, based on the LMS design, with the tare increased from 16 to 20 tons. 206 examples were built between 1951 and 1958. DB993732, at Skipton in August 1990, was built to Lot 2536 by the Birmingham Railway Carriage & Wagon Co in 1953. It is in engineer's grey and yellow and is coded Shark. Its TOPS code is ZUV.

Above: A final batch of 20 Sharks to a new Diagram, 1/598, were built by Central Wagon Co to Lot 3285 in 1960. Most accounts refer to them as being the same as their predecessors but with roller bearings and hydraulic buffers. The same underframe, 21ft over headstocks, was used, but the end verandas were 10 inches longer, the necessary savings being made by reducing the door openings from 2ft 8in to 2ft 4in, and shortening the guard's compartment by one foot to 8ft 2in. DB993935 is in weathered black livery at Smallheath on the ex-GWR Birmingham line on 1st June 1971. A faded inscription on the left hand veranda reads RETURN TO NUNEATON ABBEY ST.

Above: The Civil Engineer, with a vast fleet of ballast and rail wagons, was the most prolific user of departmental stock, but the Locomotive Department operated many interesting vehicles. The introduction of chemical water treatment plants at many LMS sheds in the 1930s and subsequently at various points on BR, led to considerable savings in boiler maintenance, but the softening process produced a repulsive white sludge. As the removal of the sludge from the softening plants was the responsibility of the Chief Mechanical Engineer's Department, they naturally used surplus equipment, and what better way to carry liquids than in old locomotive tenders! Side doors were cut in the tenders for discharge and maintenance purposes, but sludge leakage soon produced a foul mess. This former Robinson ROD tender was in the Engineer's sidings at Belper, just north of Derby, in the early 1970s. With stock sometimes moving across regional boundaries, one cannot be sure, but the most likely explanation of a Robinson tender on the LM region is that it was from one of the ROD 2-8-0s acquired after the First World War by the LNWR, and which passed to the LMS. Unlike their counterparts on the LNER, they did not have a long life, but their tenders were used for a variety of purposes.

Left: The Southern Region found another use for surplus tenders. With the premature retirement of Maunsell's superb 'Schools' class 4-4-0s in 1961-62. a number of their tenders were cut down to form the basis of snow ploughs. Numbered in the DS702xx series, DS70225 is at Salisbury on 20th July 1972.

Right: DS439 began life as Isle of Wight Railway carriage truck No 76. The Southern initially classed it as a coaching stock vehicle, numbering it 4380. In June 1930, it became a boiler trolley numbered 439S, and later DS439. A 9 foot long well was cut in the floor to accommodate locomotive fireboxes, and the trolley was used to move locomotive boilers between Ryde works and St Helen's Quay for shipment to the mainland for repairs. The wagon was strengthened with old rails and a bracket was eventually fitted to support the smokebox end of boilers. It is seen in this state at Ryde St John's works on 20th May 1964.

Centre right: Locomotive wheels sometimes needed to be transported between major sheds and works for turning. DW273, originally built on Lot 218 to Diagram 022, is a Western Region 5-plank High converted for the conveyance of loco wheels at the start of the 1950s. The floor was specially strengthened and provided with retaining chocks, visible through the open door. It carries a slot-in plate, with a 'Swindon to Banbury' inscription in classical GWR italic script and is at Banbury, in bauxite livery, on 13th September 1965, just weeks before the end of Western Region steam.

Bottom right: In the days when all but the smallest stations were staffed, and everything from small packets to train loads of minerals moved by rail, the railways required an immense number of weighing machines. To the leading makers, such as Henry Pooley or William Avery, the railways were an important source of business, not just for new machines but for servicing and repairs, as most companies left this specialised activity to the manufacturers. Repairing and calibrating weighing machines was a skilled operation requiring a good deal of equipment and a portable workshop, and what better way to take the equipment and the workshop to the station than by rail! To facilitate the work, the railways converted elderly vans for use by weighing machine contractors. This ex-Great Western van, W16547W, which still carries the old style rectangular GWR wagon plate which went out of use at the end of the 1920s, carries the inscription 'HENRY POOLEY & SON, WEIGHING MACHINE CONTRACTORS, BIRMINGHAM No 11'. In the early part of the BR era, service stock was painted black, but from March 1960 until May 1962, gulf red was used. Whilst the red paint was applied to the ends of open wagons, vans often retained black ends, as here. The next van is in weathered bauxite, giving a useful comparison between the two shades. The Weighing Machine Tool van could visit any station, however small, provided there were weighing machines to service, and would make an interesting and unusual addition to any modeller's layout.

CRANES

This section is longer than a strict sense of proportion would allow, for compared with 1.2 million wagons inherited by BR and half a million built by British Railways, rail cranes could be counted in the dozens or in three figures, so were a drop in the ocean. However strict proportion would demand that half the illustrations in this book should be of 16 ton steel mineral wagons. If justification is needed, rail cranes were colourful and interesting vehicles which have seldom been covered in colour before, and are very popular with modellers.

Below: This is my favourite crane, RS1013/50, seen at Rugby in 1964. This crane was a part of my childhood and I can recall seeing it on many visits to Rugby shed, though I never saw 1013/50 at work. It was one of the most powerful steam cranes on BR and one of the last pair built by Craven Bros for UK service in 1930. Costing the LMS over £5,000, or about one third the price of a Stanier Pacific, she was a 36 ton crane with relieving bogies and an articulated jib foot. Later 1013 was upgraded to 50 tons, as revealed by the two digit suffix to its number. One of my father's friends, the late Tom Hayward, built a superb 0 Gauge model of the whole Rugby breakdown train. It was the result of many weeks

careful measurement, and one day when the model was nearly finished, Tom noticed that the Craven roundel on his model differed from the colours on the crane. This puzzled him as he had taken careful notes. He asked shed staff what had happened. The answer was that if some paint was left over from another job, the instruction was, 'go and touch up the crane'. The Craven badge, being the most attractive item, received more than its share of repaints, but its livery changed depending on what paint happened to be left over. I offer this as a warning to modellers. We demand details of what the official livery was, but professional railwaymen were sometimes not so concerned about prototypicality!

Left: RS1076/30, seen at Rugby in August 1969, was one of five cranes built by Cowans Sheldon for the LMS in 1943. This crane is in a different livery to the one above. These are magnificent and interesting vehicles and when included on a model railway layout, they add both interest and an air of authenticity to the proceedings. I hope that the inclusion of these views in this book may in a small way encourage some modellers to resolve to recreate examples of these wonderful machines in miniature on their own railways.

BRAKE VANS

To generations of railwaymen, the idea of a freight train without a guard's van was unthinkable, but with the advent of diesel locomotives with a cab in which the guard could travel and fully air braked stock, the need for a separate van for the guard to ride in ceased and the brake van began to vanish rapidly from the railway scene. The few survivors today are normally used in connection with permanent way works trains.

Top right: As with much else on the GWR, the genesis of the Great Western 20 ton Toad brake van was laid down in the days of Churchward, and the design continued virtually unaltered until the final batch was built under a BR order in 1949. At times its detractors accuse the GWR of complacency, and there is some truth in this, but when most railways were building short wheelbase 10 tonners, Swindon had seen the benefits of the smoother riding long wheelbase 20 ton van, and saw no need to alter the design thereafter. By 1973, when W17441, was photographed at Shepherd's Well, the junction of the stump of the East Kent Railway with the former SECR main line between Dover and Canterbury, the Toad was a rarity, for it had been banned from main line use following union pressure. The reason was that unlike the majority of later designs, the brake wheel was on the open veranda, rather than inside the van, making them less pleasant to operate in winter. W17441 and a sister van W17480, were no longer used as guard's vans, and had lost their lower full length step boards, but had found a use far from Great Western metals providing a fitted brake head on the Tilmanstone coal workings.

Centre right: The Southern Railway inherited a variety of brake vans from its constituent companies, many dating back to the 1870s or 1880s. Most perished as a part of SR modernisation plans, but a few found a fresh lease of life in the Isle of Wight. S56058, at Ryde St John's Road in October 1966, was an ex-LSWR Diagram 1542 Large Road Van, of which four were built in 1906. It was transferred to the Island in 1947 and, with its 14 foot wheelbase, was the largest on the Island. It was condemned in 1966. The second van, S548, is a Diagram 1541 Standard Road Van with a 10ft 6in wheelbase.

Bottom right: As with the GWR, the Southern was aware of the improved riding characteristics of a longer wheelbase, but preferred a shorter body with verandas at each end. First built in the mid-'twenties, the new SR standard van, which was based upon an SECR designed chassis, was produced for the rest of the company's existence. Though No 55167 was completed in 1949, it has been restored in Southern colours by the North Norfolk Railway. It is depicted at Sheringham on 14th June 1975.

Top: The Southern handled the least volume of freight of the Big Four, so it is perhaps surprising that it produced the most sophisticated brake vans ever to operate in freight service in England. Following the successful rebuild of the chassis of some old electric stock as bogie vans, the SR produced some new vans in 1936. A short body was mounted on a robust bogie chassis, and the combination of long wheelbase, coach type bogies and plenty of weight produced a vehicle with splendid riding characteristics. After their revenue earning days were over, many were transferred to departmental duties. S56288 is seen with ELECTRIFICATION lettering and striped ends, but otherwise still in traffic livery, at Rugby maintenance depot on 21st June 1980.

Above: As might be expected, the LMS was heavily influenced by Midland thinking, and the final MR brake van design became the first standard LMS design too, with over 800 examples being built between 1924 and 1927. DM1267, branded for use as a crane mess van and seen at Horwich on 16th August 1980, was 20ft over headstocks with a 12ft wheelbase. It is in departmental stock green livery with wasp stripe ends. *Chris Nettleship*

Top right: After building many hundreds of shorter vans and a transitional type, the LMS caught up with the GWR with its Diagram D1919 20 ton goods brake vans, of which 670 were built at Derby between 1935 and 1938. M730302, at Wansford in the early days of the Nene Valley Railway in 1976, is still in BR livery and gives a good idea of how traffic vehicles were patch painted when put into departmental use.

Below right: At first sight a similar vehicle to M730302, but close study reveals that DM731411, at Windermere on 19th September 1971, is to the final LMS Diagram D2068 design. The lookout ducket is 3 inches wider than on the earlier design, whilst the weight boxes are much deeper. Vehicle details appear in the post-1963 boxed style with the departmental 'D' added subsequently. Modellers should note the grey-brown running gear and weight boxes, the dingy state of the body paintwork and the frayed canvas roofing on the near end.

Bottom: The LNER introduced some trial long wheelbase vans in 1929, initially confining them to long distance fast freights. The new 20 ton design, with a short body and long end platforms, was even more long lived than its Southern counterpart, as it was adopted with few modifications as the BR standard van. Early LNER examples had cast iron weights to provide added brake power, but during the war, concrete weights were used to save steel stocks. At first sight, BR vans, such as B951270, were identical to the LNER design, but closer inspection reveals that the footboards, instead of stopping just beyond the axleboxes, extend the full length of the chassis. This Diagram 1/504 van, built at Faverdale wagon shops in 1951 to Lot 2206, is seen at Birkenhead Docks in 1982. The locomotive, 03 170, was one of the last trio of 03 shunters active on mainland British Rail.

How better to end this part of the book than with a classic portrait of the guard keeping a sharp lookout from his brake van, the traditional end of a freight train from the dawn of railways until the phasing out of hand braked freight stock in the years after the end of steam traction on Britain's railway network in 1968. B954270, a Diagram 1/506 brake van, was built at Faverdale to Lot 3129 in 1958 and was photographed, in glorious sunshine, on the down goods line at Rugby on 16th March 1971. At Rugby, the LNWR built a massive train shed which spanned the platforms and fast roads, but on the down side, a horse and carriage landing meant that the down goods ran outside the train shed at the London end, but curved back under the train shed through a massive opening at approximately the mid-point of the station. Unfitted freight trains carried side lamps as well as central tail lamps. In order that the engine crew could confirm that the train was complete, the side lamps displayed white aspects forwards. Rule 121 of the 1950 BR rule book stated that whilst passenger and fully fitted freights displayed one red light to the rear, other freights showed two red lights on main lines, fast lines and single lines, but when running on slow, relief or goods lines adjacent to running lines, the lamp nearest the main line was to show a white aspect, giving one red and one white. On goods lines adjacent to slow lines, trains were to carry two red lights.

Appendix A

HEADCODES
& BLOCK CODES

In 1960, when new British Railways' General Appendices and Block Regulations were introduced, traditional headlamp codes and block codes still prevailed. By 1972 when a new set of instructions were issued, the system had changed almost out of recognition with old train types vanishing and new ones appearing. As the number of practical headcodes and bellcodes was limited, old friends sometimes appeared in unexpected new guises.

The unfitted coal train of 1960 had vanished in 1962, but by 1971 its bellcode heralded the fully fitted express freight, and its headcode adorned the empty coaching stock train!

This period of rapid change, with revisions in 1962, 1968, 1969 and 1971, has confused enthusiasts and even professional railway staff. In the next three pages we cover the 1960, 1962 and 1971-72 versions.

1960 STANDARD CODE OF ENGINE HEAD LAMPS OR DISCS, AND BLOCK TELEGRAPH CODES

Classifi-cation	Engine Headcode (white lights or discs)	Description of Train	Block Telegraph Codes
A		Express passenger train, newspaper train or breakdown van train or snow plough going to clear the line or light engine going to assist disabled train.	4
		Officers' Special train *not* requiring to stop in section.	–
B		Ordinary passenger train, mixed train or breakdown van train *not* going to clear the line	3 – 1
		Branch passenger train. (To be used only where authorised by the regional Operating Officer.)	1 – 3
C		Parcels, fish, fruit, horse, livestock, meat, milk, pigeon or perishable train composed entirely of vehicles conforming to coaching stock requirements.	1 – 3 – 1
		Express freight, livestock, perishable or ballast train piped fitted throughout with the automatic brake operative on not less than half of the vehicles.	3 – 1 – 1
		Empty coaching stock train (not specially authorised to carry 'A' headcode).	2 – 2 – 1
D		Express freight, livestock, perishable or ballast train partly fitted with the automatic brake operative on not less than one third of the vehicles.	5
E		Express freight, livestock, perishable or ballast train partly fitted, with not less than four braked vehicles next to the engine and connected by the automatic brake pipe.	1 – 2 – 2
		Express freight, livestock, perishable or ballast train with a limited load of vehicles, *not* fitted with the automatic brake.	1 – 2 – 2
F		Express freight, livestock, perishable or ballast train *not* fitted with the automatic brake.	3 – 2
G		Light engine or light engines coupled.	2 – 3
		Engine with not more than two brake vans.	1 – 1 – 3
H		Through freight or ballast train *not* running under class 'C', 'D', 'E' or 'F' headcode	1 – 4
J		Mineral or empty wagon train.	4 – 1
K		Freight, mineral or ballast train stopping at intermediate stations.	3
		Branch freight train. (To be used only where authorised by the regional Operating Officer.)	1 – 2
		Freight, ballast or Officers' Special train requiring to stop in section.	2 – 2 – 3

Notes: 1. The above arrangements do not apply on the Southern Region.

2. When a train is worked by two engines attached in front of the train, the second engine must not carry head lamps or discs.

1962 STANDARD CLASSIFICATION AND CODE OF HEAD LAMPS OR DISCS, AND BLOCK TELEGRAPH CODES

Classification (first frame of indicator box)	Engine Headcode (white lights or discs)	Description of Train	Block Telegraph Codes
1		Express passenger train, newspaper train or breakdown van train or snow plough going to clear the line or light engine going to assist disabled train.	4
		Officers' Special train *not* requiring to stop in section.	–
2		Ordinary passenger train, mixed train or breakdown van train *not* going to clear the line	3 – 1
		Branch passenger train. (To be used only where authorised by the Regional Operating Officer.)	1 – 3
3		Parcels, fish, fruit, horse, livestock, meat, milk, pigeon or perishables train composed entirely of vehicles conforming to coaching stock requirements.	1 – 3 – 1
		Empty coaching stock train (*not* specially authorised to carry '1' headcode).	2 – 2 – 1
4		Express freight train pipe fitted throughout with the automatic brake operative on *not* less than 90% of the vehicles. Maximum speed 55 mph. *	3 – 1 – 1
5		Express freight train partly fitted, with the automatic brake operative on *not* less than half of the vehicles. Maximum speed 50 mph.	5
6		Express freight train partly fitted, with the automatic brake operative on *not* less than 20% of the vehicles. Maximum speed 45 mph.	1 – 2 – 2
7		Express freight train *not* fitted with the automatic brake. † Maximum speed 40 mph.	3 – 2
8		Through freight train *not* fitted with the automatic brake. † Maximum speed 35 mph.	1 – 4
9		Branch train or stopping freight train and Officers' Special train or ballast train requiring to stop in section. † Maximum speed 35 mph.	Stopping freight 3 Branch train 1 – 2 Officers' Special 2 – 2 – 3
0		Light engine or light engines coupled. ‡	2 – 3
		Engine with not more than two brake vans. ‡	1 – 1 – 3

Notes:

1 The above arrangements do not apply on the Southern Region, except where specially authorised.

2 Empty wagons and ballast trains should run at the highest classification appropriate to the braked portion available and the type of wagon conveyed.

* A maximum speed of 60 mph will apply in respect of certain trains specifically indicated in the Working Timetables.

† Where fitted vehicles are required to be placed next to the engine and coupled up, this will be shewn in the appropriate Regional train loading instructions.

‡ Maximum speed according to class of engine and type of brake van.

1971/72 STANDARD CLASSIFICATION AND CODE OF HEAD LAMPS OR DISCS, AND BLOCK TELEGRAPH CODES

Classification (first frame of indicator box)	Engine Headcode (white lights or discs)	Description of Train	Block Telegraph Codes
1	(two lamps, bottom left & bottom right)	Express passenger train, postal train, newspaper train, or breakdown van train or snow plough going to clear the line, or light locomotive going to assist disabled train.	4
		Officers' Special train *not* requiring to stop in section.	–
2	(one lamp, top centre)	Ordinary passenger train, mixed train or breakdown van train or snow plough *not* going to clear the line	3 – 1
3	(two lamps, bottom left & bottom right)	Express parcels train composed of vehicles permitted to run at 90 mph or over.	1 – 3 – 1
4	(two lamps, top centre & bottom centre)	Freightliner train.	3 – 2 – 5
		Parcels train, Company or express freight train composed of vehicles permitted to run at 75mph or over.	3 – 1 – 1
5	(one lamp, bottom left)	Empty coaching stock train (not specially authorised to carry Class '1' headcode.	2 – 2 – 1
6	(two lamps, top centre & bottom left)	(a) Fully fitted Company or block train, parcels train or milk train.	5
		(b) Ordinary fully-fitted express freight train, with brake force not less than that shown in Section E of the Loads Book.	4 – 1
7	(two lamps, bottom left & bottom right)	Express freight train, not fully-fitted but with brake force not less than that shown in Section E of the Loads Book.	1 – 2 – 2
8	(two lamps, top centre & bottom right)	Freight train, not fully-fitted but with brake force not less than that shown in Section E of the Loads Book.	3 – 2
9	(one lamp, bottom right)	Unfitted freight train (where specially authorised).	1 – 4
		Freight train, Officers' Special train or Engineers' train requiring to stop in section..	2 – 2 – 3
0	(one lamp, bottom centre)	Light locomotive, light locomotives coupled, or locomotive with brake tender(s).	2 – 3
		Locomotive with not more than two brake vans.	1 – 1 – 3

Notes:

1. Trains in Class 6(a) will be timed according to the maximum speed of the vehicles scheduled to be conveyed.
2. Trains in Class 6(b), 7, 8 and 9 will be timed to reflect a maximum speed of 45 mph, or such other lower maximum speed it may be necessary to impose on individual trains.
3. Fully fitted train: – A train with all the vehicles fitted with the automatic brake or brake-pipe coupled up and in use.

    A brakevan will not normally be provided on a parcels or fully-fitted freight train, and when not provided the last two vehicles must be fully-fitted with the automatic brake in working order. I

    If circumstances require a brakevan to be provided on a fully-fitted freight train it must be marshalled at the rear and the Guard must ride in it. The brakevan may be piped only.

PAINTING GUIDE FOR STOCK CONVEYING DANGEROUS GOODS

Information seldom appears in enthusiast literature about the painting of wagons conveying explosives or inflammable liquids. When building chemical tankers for our own layout, I was able to refer back to RCH or BR official handbooks which few modellers will have access to, so rather than include the usual general livery table, which would repeat information available elsewhere, I have prepared this summary from the July 1957 issue of the British Transport Commission regulations for the conveyance of dangerous goods by merchandise trains. The BTC listed nine categories of hazardous merchandise.

Section 1
EXPLOSIVES

In pre-grouping days, gunpowder vans were painted conspicuously and prominently lettered. In the grouping era, they became less conspicuous, and this trend continued in BR days. Standard unfitted or fitted stock liveries were applied, and only their smaller size, the word GUNPOWDER on the wagon, and a small instruction plate distinguished them from ordinary stock. Gunpowder vans were to be used for explosives, but NOT for other traffics unless specially authorised. Not more than five vehicles containing explosives were to be loaded or unloaded at any goods depot at any one time, nor were more than five wagons to run in any train. Separate rules applied to explosives conveyed for the Forces.

Section 2
COMPRESSED OR LIQUIFIED GASES

The barrels of tank wagons were semi-gloss or gloss white, with a band of signal red, (BS colour 537) 6 inches wide, round each end of the barrel, the outer edge of the band being between 6 and 12 inches from the end. Owners could paint other coloured bands at their discretion, but these had no railway significance, and were placed centrally, or round the dome of the tank. A steel plate, preferably finished in vitreous enamel, was fitted vertically on top of each solebar at the left hand end to carry the name of the gas, statutory and common sense warnings, and an enquiry address. For gases where the main hazard was fire, white lettering on a red background, was specified. For other gases, the lettering was black on golden yellow (BS 556). Solebars were black or dark grey. Typical substances conveyed in these wagons included nitrogen, oxygen, hydrogen, anhydrous ammonia, chlorine and propane.

Section 3
CORROSIVES

The barrels of tank wagons could be any colour other than white or aluminium. Solebars could NOT be painted red. Substances in this category included most types of acid such as sulphuric, nitric and hydrochloric acids and many chlorides.

Section 4
POISONS

The barrels of tank wagons could be any colour other than white or aluminium. Solebars could NOT be painted red. Tank wagons carried a solebar plate giving the name of the commodity and the word POISON, with an agreed warning notice. Substances included weedkillers and motor fuel anti-knock compounds containing Lead Tetra Ethyl.

Section 5A
HIGHLY INFLAMMABLE LIQUIDS

The main hazard was fire or an explosion caused by the ignition of a a mixture of inflammable vapour and air. This applied to liquids having a closed flash point below 73 degrees fahrenheit. Tank barrels were painted aluminium with agreed inscriptions, and solebars were signal red. Light grey was later substituted for aluminium. Tank wagons were not to be filled or emptied on railway premises except at approved sites. Substances in this classification included benzene, industrial alcohol, ethanol, propanol, petrol and petroleum ether.

Section 5B
INFLAMMABLE LIQUIDS

These were defined as liquids with a flash point between 73 and 150 degrees fahrenheit. Tank barrels could be any colour other than white or aluminium. Solebars were not to be red. In practice, class B tankers were invariably painted black. Substances included coal tar, oil, fuel oil, gas oil, kerosene and shale oil.

Section 6
SUBSTANCES WHICH REACT VIOLENTLY TO WATER

The potential hazard was inflammable or poisonous gases. Such substances were to be conveyed in waterproof sacks, wooden casks, drums, suitably protected bottles, or in bulk in approved watertight wagons. They included alkalis, Calcium Carbide or Magnesium powder.

Section 7
STRONG SUPPORTERS OF COMBUSTION

Tank wagons for Hydrogen Peroxide solution, of not more than 40 per cent strength, were to have barrels any colour other than white or aluminium. Solebars were NOT to be red.

Section 8
SPONTANEOUSLY COMBUSTIBLE SUBSTANCES

Tank wagons containing phosphorus had barrels painted aluminium, and black solebars, with approved warning notices on the barrel or solebar plates. Demountable tanks were to be aluminium with approved warning notices. The combination of an aluminium barrel and black solebars appears to be unique to yellow or white phosphorus, as it could heat spontaneously to a degree which would cause a fire, or ignite spontaneously in air.

Section 9
RADIO ACTIVE SUBSTANCES

The movement of radio active substances by rail was in its infancy in 1957 and was subject to special arrangements. The standard regulations ran to just five lines of print at that time!

'Radioactive substances are accepted only by special arrangement, except those in quantity not exceeding 0.1 millicurie or articles such as clocks or instruments having luminous dials or other sources of radiation packed so that the radiation at any surface of the package shall not exceed 10 milliroentgens per 24 hours, which are regarded as non-dangerous'.

The BR wagon diagram books were large volumes containing some hundreds of outline wagon drawings. We can only offer a representative selection here. Their main purpose was to meet the needs of the commercial department, so detail below solebar level was lacking. With such a wide range of ready to run stock or kits available in all major scales nowadays, most modellers tend to adapt proprietary stock or use a commercial chassis and a scratch built body so the lack of this detail should not be a problem.

Wagon diagram books were produced by the most economical means available to provide basic data at a very low cost. Blurred or lost figures would often appear on every copy of a given diagram. In some cases in the pages that follow, diagrams have been included which contain these minor faults.

11 TON GUNPOWDER VAN

Diagram 1/261 was for 15 11 ton Gunpowder vans (B887120-887134) built to Lot 3099 at Swindon in 1958. A further 25 wagons were added under Lot 3237 in 1959. They were a development of the Diagram 1/260 BR Gunpowder van of which 120 had been built between 1953 and 1956, the most significant change being the adoption of a 10ft wheelbase. Both designs followed GWR practice. This combination of an elderly pre-nationalisation design with some minor improvements typified many BR wagon diagrams.

LOT 3099 QTY. 15   261

Scale: 4mm = 1 ft

TEL. CODE – CONE.
BRAKE – A.V.B. & .HAND.
CARRYING CAPACITY – 11 TON.
TARE – 8 TON 1 CWT.
MINIMUM CURVE – 1 CHAIN.

BODY – STEEL (WOOD LINED)
FLOOR – WOOD.

WHEELS – 3'-1½" DIA.
JOURNALS – 9" x 4¼"
BUFFERS – HYDRAULIC
BUFFER HEAD – 1'-1" DIA.
COUPLINGS – CONTINENTAL SCREW.

20 TON GLW COVERED GOODS VAN

In 1970 40 vacuum braked wagons were rebuilt with air brakes to test their suitability for use in the new air braked fleet.
After trials, other stock was subsequently rebuilt, including Bogie Bolsters and Pipes. Diagram 1/263 covered the rebuilding of six Diagram 1/211 Palvans. The drawing, which would be suitable for unrebuilt Palvans to the original diagram, or the rebuilds, shows the large off-set door and cross bracing on the sides.

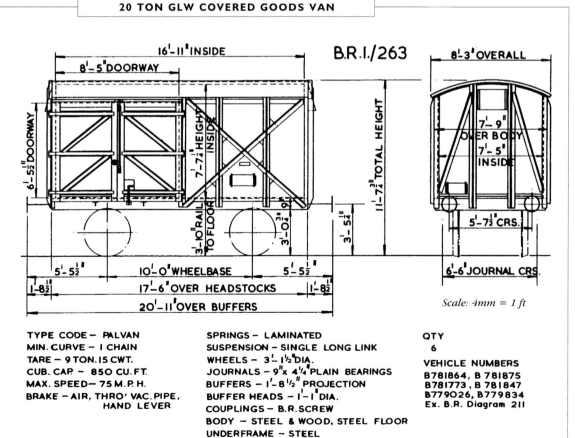

B.R.I./263

Scale: 4mm = 1 ft

TYPE CODE – PALVAN
MIN. CURVE – 1 CHAIN
TARE – 9 TON. 15 CWT.
CUB. CAP. – 850 CU. FT.
MAX. SPEED – 75 M.P.H.
BRAKE – AIR, THRO' VAC. PIPE, HAND LEVER

SPRINGS – LAMINATED
SUSPENSION – SINGLE LONG LINK
WHEELS – 3'-1½" DIA.
JOURNALS – 9" x 4¼" PLAIN BEARINGS
BUFFERS – 1'-8½" PROJECTION
BUFFER HEADS – 1'-1" DIA.
COUPLINGS – B.R. SCREW
BODY – STEEL & WOOD, STEEL FLOOR
UNDERFRAME – STEEL

QTY 6

VEHICLE NUMBERS
B781864, B 781875
B781773, B 781847
B779026, B779834
Ex. B.R. Diagram 211

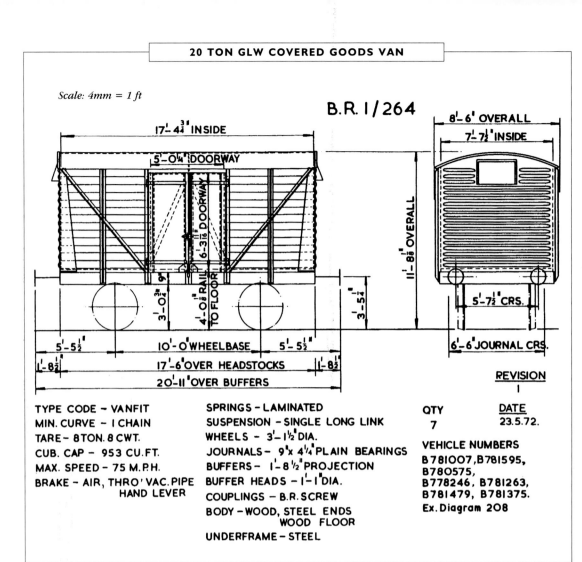

20 TON GLW COVERED GOODS VAN

Scale: 4mm = 1 ft

B.R. I/264

17'-4¾" INSIDE

5'-0¼" DOORWAY

6'-3⁵⁄₁₆" DOORWAY

8'-6' OVERALL

7'-7½" INSIDE

11'-8⅛" OVERALL

4'-0¼" RAIL TO FLOOR

3'-0¾"

3'-5¼"

5'-7½" CRS.

5'-5½"

10'-0" WHEELBASE

5'-5½"

6'-6" JOURNAL CRS.

1'-8½"

17'-6" OVER HEADSTOCKS

1'-8½"

20'-11" OVER BUFFERS

REVISION
I

QTY   DATE
7   23.5.72.

TYPE CODE – VANFIT
MIN. CURVE – I CHAIN
TARE – 8 TON. 8 CWT.
CUB. CAP – 953 CU. FT.
MAX. SPEED – 75 M.P.H.
BRAKE – AIR, THRO' VAC. PIPE
        HAND LEVER

SPRINGS – LAMINATED
SUSPENSION – SINGLE LONG LINK
WHEELS – 3'-1½" DIA.
JOURNALS – 9" x 4¼" PLAIN BEARINGS
BUFFERS – 1'-8½" PROJECTION
BUFFER HEADS – 1'-1" DIA.
COUPLINGS – B.R. SCREW
BODY – WOOD, STEEL ENDS
            WOOD FLOOR
UNDERFRAME – STEEL

VEHICLE NUMBERS
B781007, B781595,
B780575,
B778246, B781263,
B781479, B781375.
Ex. Diagram 208

Diagram 1/264 was issued as a part of the same process, as it covered the rebuilding of seven Diagram 1/208 vans with airbrakes and vacuum through pipes. The Diagram 1/208 van was the definitive BR van, of which over 19,000 were built between 1951 and 1958. As with Diagram 1/263, it gives the running numbers of the rebuilds, and the modeller who enjoys an unusual prototype could adapt a standard Diagram 1/208 model to Diagram 1/264, and run it in an airbraked freight, confounding all but the most discerning critics!

The Diagram 1/270 20 ton Grain Hopper, was a re-issue of the LMS Diagram D1689 Grain van which dated from 1928, but with the more rounded top characteristic of later LMS deliveries. Numbered B885000-885039, 40 were built to Lot 2009 at Derby in 1949, but from 1951, a revised design, omitting the end portholes, was produced.

20 TON HOPPER GRAIN VAN

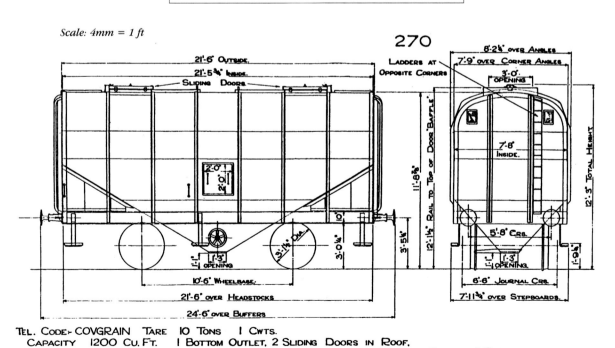

Scale: 4mm = 1 ft

270

21'-6" OUTSIDE.

21'-5¾" INSIDE.

SLIDING DOORS

6'-2¼" OVER ANGLES

7'-9" OVER CORNER ANGLES

3'-0" OPENING

LADDERS AT OPPOSITE CORNERS

11'-8⅝" 12'-1½" RAIL TO TOP OF DOOR BAFFLE

7'-8" INSIDE.

12'-3" TOTAL HEIGHT

2'-0" 2'-0"

3'-1½" DIA.

3'-0¾"

3'-5¼"

10'-6" WHEELBASE.

1'-1" 1'-3" OPENING

5'-8" CRS.

1'-9¾"

1'-1" 1'-3" OPENING

21'-6" OVER HEADSTOCKS

6'-6" JOURNAL CRS.

7'-11¾" OVER STEPBOARDS.

24'-6" OVER BUFFERS

TEL. CODE – COVGRAIN   TARE 10 TONS I CWTS.
CAPACITY 1200 CU. FT.   I BOTTOM OUTLET. 2 SLIDING DOORS IN ROOF.
I SIDE TRAP DOOR. INDEPENDENT BRAKE EACH END.   MINIMUM CURVE I CHAIN

Scale: 4mm = 1 ft

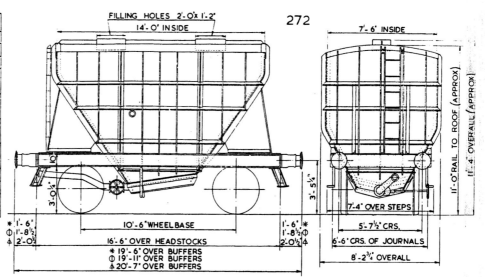

COUPLING	BUFFER PROJ.	LOT. No.	QTY.	TARE T - CWT
INSTANTER	1'- 6"	2769	110	12 - 5
INSTANTER	1'- 6"	2863	70	12 - 5
STD. SCREW	1'- 8½"	3029	80	12 - 14
CONT. SCREW	2'- 0½"	3156	50	13 - 0
STD. SCREW	1'- 8½"		220	
CONT. SCREW	2'- 0½"	3175	30	13 - 2
STD. SCREW	1'- 8½"		270	
CONT. SCREW	2'- 0½"	3176	26	13 - 2
STD. SCREW	1'- 8½"		74	
CONT. SCREW	2'- 0½"	3177	80	13 - 1
STD. SCREW	1'- 8½"		120	
INSTANTER	1'- 8½"	3323	170	12 - 18
INSTANTER	1'- 8½"	3361	170	12 - 18
INSTANTER	1'- 8½"	3406	150	12 - 18
INSTANTER	1'- 8½"	3409	150	12 - 19

TEL CODE - PRESFLO
CUBIC CAPACITY - 611 CU. FT.
MINIMUM CURVE - 1 CHAIN
A.V.B. & HAND BRAKE

WHEELS - 3'- 1½" DIA.
JOURNALS - 10"x 5" OR 4⅞" DIA. ROLLER BEARINGS
BUFFER HEADS - 1'- 1" DIA.

Unlike the Diagram 1/270 Grain hopper, the Diagram 1/272 Hopper Cement Wagon or Presflo owed nothing to pre-nationalisation designs. The first examples appeared in 1955, and over the next few years batches appeared with Instanter, Standard Screw or Continental Screw couplings and buffers to suit, though the bodies were standard in all cases. The diagram is of particular interest as it shows the non-symmetrical nature of the design, with the discharge pipe and access ladder at one end only. The 10ft 6in wheelbase is unusual for a 16ft 6 in chassis, but is explained by the need to accommodate the lower portion of the hopper.

Scale: 4mm = 1 ft

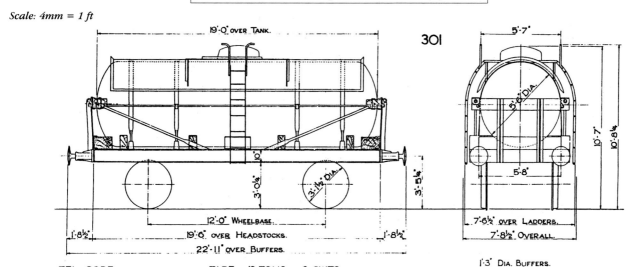

TEL. CODE.
CAPACITY OF TANK 2490 GALLS.
STD. 20 TON LEVER BRAKE. THRO' PIPE

TARE 12 TONS 9 CWTS

BR owned very few tank wagons, less than 200 being constructed to all BR designs. The reason for this diminutive total was that most petro-chemical traffic was moved in privately owned stock. Most of the handful of BR tanks were not even for petro-chemical traffic, but for other liquids, but one of the exceptions was Diagram 1/301, which was for 10 Ethylene Oxide tankers, B749600-749609, built by Grazebrook in 1955 under Lot 2170. They would have been hired out to industrial users. With their long (12 foot) wheelbase, they were typical of contemporary PO stock.

Scale: 4mm = 1 ft

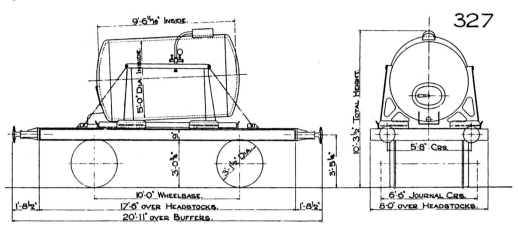

327

TEL. CODE-   TARE- 7 TONS 13 CWTS.
CAPACITY OF TANK 1080 GALLS.
A.V.B. & HAND BRAKE. THRO' STEAM PIPE.
MINIMUM CURVE 1 CHAIN.

Diagrams 1/325 to 1/343 were for demountable beer tanks, Diagram 1/327 being for ten wagons, B749023-749032, built at Derby under Lot 2073 in 1949. Shortly before commencing work on this book, I scratchbuilt one of the demountable tank wagons in 0 Gauge, using a standard 17ft 6in underframe, the tank itself coming from a suitably reshaped 4mm body-only moulding for a modern Monobloc tanker.

This provided an interesting and unusual wagon which has generated considerable interest.

8 & 12 TON CATTLE WAGONS

Scale: 4mm = 1 ft

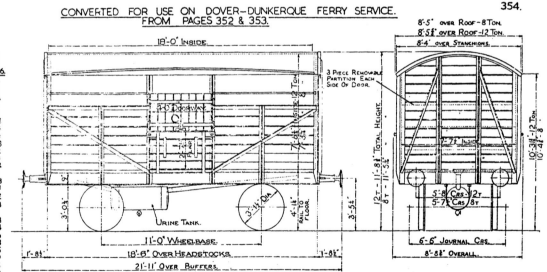

CONVERTED FOR USE ON DOVER-DUNKERQUE FERRY SERVICE.
FROM PAGES 352 & 353.

354.

VEHICLE NOS.	
8 TON.	12 TON.
QUAN. 14.	QUAN. 26.
B893083	B891509
" 3107	" 1636
" 3162	" 1644
" 3174	" 1677
" 3191	" 1679
" 3398	" 1703
" 3409	" 1759
" 3429	" 1799
" 3430	" 1808
" 3431	" 1814
" 3504	" 1849
" 3538	" 1904
" 3560	" 1951
" 3542	" 2033
	" 2061
	" 2132
	" 2133
	" 2181
	" 2253
	" 2264
	" 2296
	" 2298
	" 2325
	" 2469
	" 2475
	" 6170

TEL. CODE - OXFIT.   TARE - 8 TON - 4 CWT - 12T WAGONS.
                  " 7 " 18 " - 8 "   "
A.V.B. & HANDBRAKE.   DOORS.- 2 SIDE & 2 FLAP.
MIN. CURVE - 1 CHAIN.   FLOOR AREA - 137 SQ. FT.
CU. CAP. -12T. = 955 CU. FT. ----8T. = 920 CU. FT.

JOURNAL SIZE - 8 TON = 8"x3¾" -- 12TON = 9"x4¼"
BUFFERS- SPINDLE.   BUFFER HEADS - 1"x1" DIA.
BODY - WOOD.   FLOOR - WOOD.
COUPLINGS - SCREW.

The Diagram 1/354 cattle wagon was a modification of the 12 or 8 ton cattle vans built to Diagrams 1/352 and 1/353 to suit them for the Dunkerque rail ferry service. It shows the positioning of the urine tank, and in giving dimensions for both types of wagon, reveals how absurd it was to have 8 and 12 ton vehicles which were all but indistinguishable. The numbers of the individual wagons which were converted appear on the left of the diagram.

Scale: 4mm = 1 ft

390.

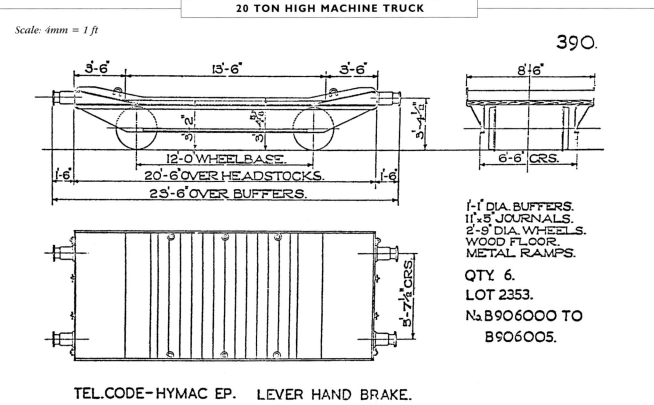

3'-6"   13'-6"   3'-6"

8'-6"

3'-1½"

12'-0" WHEELBASE.
20'-6" OVER HEADSTOCKS.
23'-6" OVER BUFFERS.

6'-6" CRS.

8'-7½" CRS.

1-1" DIA. BUFFERS.
11"x5" JOURNALS.
2'-9" DIA. WHEELS.
WOOD FLOOR.
METAL RAMPS.

QTY. 6.
LOT 2353.
Nº B906000 TO
B906005.

TEL. CODE—HYMAC EP.   LEVER HAND BRAKE.
TARE— 7 TONS 3 CWT.   MINIMUM CURVE-1¼ CHAINS.

The Hymac, or High Machine Trolley was another rare wagon, with just 4 BR Diagrams, 2/390 to 2/393, with 27 examples in total. The earliest were 6 wagons, B906000-906005, built by Lancing works under Lot 2353 in 1952.

Given the existence of the Lowmac for tall vehicles, and the single plank Lowfit for smaller loads, it is hard to see what niche the Hymac filled, and what is puzzling is not why so few were built, but why any were built at all!

Scale: 4mm = 1 ft

416

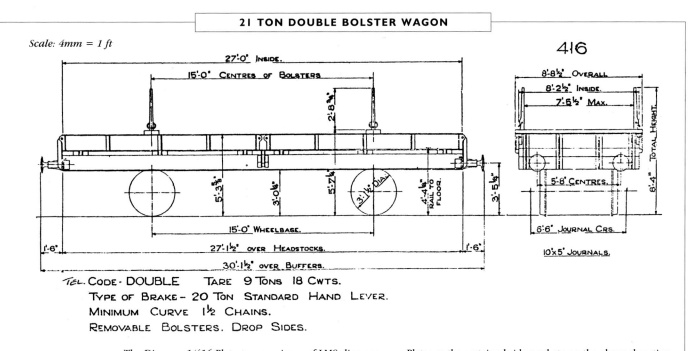

27'-0" INSIDE.
15'-0" CENTRES OF BOLSTERS
2'-8¾"

8'-8½" OVERALL
8'-2½" INSIDE
7'-6½" MAX.

5'-3⅝"   3'-0⅜"   5'-7¼"   3'-1½" DIA.   4'-4½" RAIL TO FLOOR

3'-5¼"

8'-4" TOTAL HEIGHT

15'-0" WHEELBASE.
27'-1½" OVER HEADSTOCKS.
30'-1½" OVER BUFFERS.

5'-8" CENTRES.
6'-6" JOURNAL CRS.
10"x5" JOURNALS.

TEL. CODE - DOUBLE   TARE 9 TONS 18 CWTS.
TYPE OF BRAKE - 20 TON STANDARD HAND LEVER.
MINIMUM CURVE 1½ CHAINS.
REMOVABLE BOLSTERS. DROP SIDES.

The Diagram 1/416 Plate was a re-issue of LMS diagram D2105, 200 being turned out from Wolverton in 1949. By the mid-'fifties the need for single or even Double Bolsters was falling and many were converted to Plate wagons. The conversions could be distinguished from the genuine

Plates as they retained side pockets on the channel section solebar beneath each bolster stanchion. Regrettably these do not appear on the diagram, for as I have already commented, the diagram books were for the commercial department and seldom gave much underframe detail.

Scale: 4mm = 1 ft

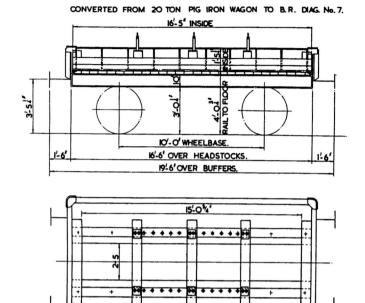

CONVERTED FROM 20 TON PIG IRON WAGON TO B.R. DIAG. No. 7.

16'-5" INSIDE

3'-5½"

3'-0½"

4'-0¾" RAIL TO FLOOR

10'-0" WHEELBASE.

16'-6" OVER HEADSTOCKS.

1'-6'

19'-6' OVER BUFFERS.

15'-0¾'

2'-5'

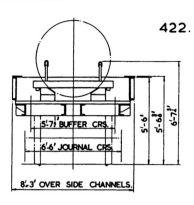

422.

5'-7½" BUFFER CRS.

6'-6' JOURNAL CRS.

5'-6' / 5'-6½' / 6'-7¾'

8'-3' OVER SIDE CHANNELS.

BRAKE:- HAND CLASP TYPE.
TARE:- 9 TONS 6 CWTS.
WHEELS:- 3'-1½' DIA.
JOURNALS:- 9'x5'.
BUFFERS:- SELF CONTAINED.
BUFFER HEADS:- 1'-1' DIA.
COUPLINGS:- THREE LINK.

The Diagram 1/422 strip coil wagon was a conversion of the Diagram 1/007 twenty ton pig iron wagon, 100 of which, B744780-744879, were built to Lot 3085 by Derby in 1958.

A number were converted to Coil C after a few years, with a wooden framework with adjustable cross pieces fitted inside the existing body.

Scale: 4mm = 1 ft

434

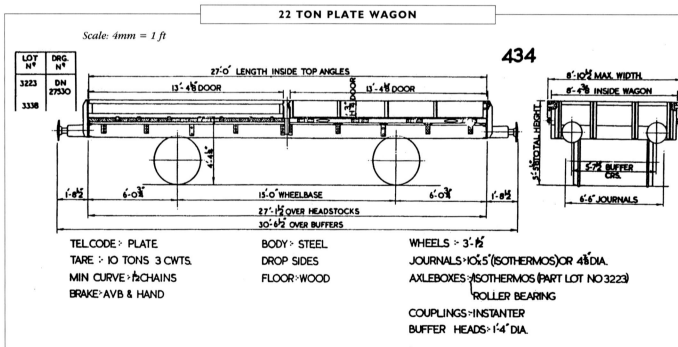

LOT Nº	DRG. Nº
3223	DN 27330
3338	

27'-0' LENGTH INSIDE TOP ANGLES

13'-4⅝' DOOR

13'-4⅝' DOOR

4'-4⅝'

1'-8½'

6'-0¾'

15'-0' WHEELBASE

6'-0¾'

1'-8½'

27'-1½' OVER HEADSTOCKS

30'-6½' OVER BUFFERS

8'-10½' MAX. WIDTH

8'-4⅝' INSIDE WAGON

5'-5' TOTAL HEIGHT

5'-7½' BUFFER CRS.

6'-6' JOURNALS

TEL CODE:- PLATE
TARE:- 10 TONS 3 CWTS.
MIN CURVE:- 1½ CHAINS
BRAKE:- AVB & HAND

BODY:- STEEL
DROP SIDES
FLOOR:- WOOD

WHEELS:- 3'-1½'
JOURNALS:- 10'x5' (ISOTHERMOS) OR 4⅛' DIA.
AXLEBOXES:- ISOTHERMOS (PART LOT NO 3223)
ROLLER BEARING
COUPLINGS:- INSTANTER
BUFFER HEADS:- 1'-4' DIA.

The LMS and LNER had built almost identical Double Bolsters in their last years and the same situation prevailed with Plate wagons. BR Diagram 1/430 was a reissue of an LMS diagram, whilst Diagram 1/431 was the LNER counterpart. The early BR Plates were unfitted, but in 1956-58, 1,200 vacuum fitted Plates were built to Diagram 1/432. These were followed by a further 2,500 fitted wagons to Diagram 1/434 in 1959-61, the only significant difference to the previous batch being BR clasp brakes instead of LNER type brakes, as on Diagram 1/432.

The building of so many Plates at such a late date in the standard wagon era shows how rapidly this traffic was expanding. However, to issue four diagrams for Plates reflects the muddled and inconsistent nature of BR policy, for trivial differences merited separate diagrams in a fleet of less than 7,000 Plate wagons, whilst major differences were tolerated in the Diagram 1/108 steel minerals or Diagram 1/208 goods van without the introduction of a separate diagram.

22 TON TUBE WAGON

Scale: 3.5mm = 1 ft

B.R.1/448

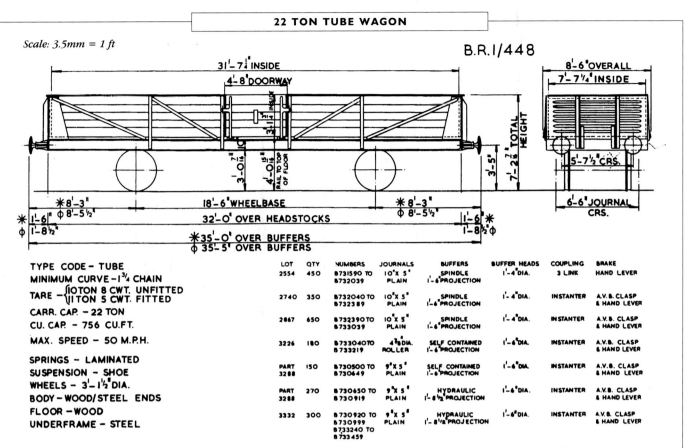

TYPE CODE – TUBE
MINIMUM CURVE – 1¾ CHAIN
TARE – {10 TON 8 CWT. UNFITTED / 11 TON 5 CWT. FITTED}
CARR. CAP. – 22 TON
CU. CAP. – 756 CU.FT.
MAX. SPEED – 50 M.P.H.

SPRINGS – LAMINATED
SUSPENSION – SHOE
WHEELS – 3'-1½" DIA.
BODY – WOOD/STEEL ENDS
FLOOR – WOOD
UNDERFRAME – STEEL

LOT	QTY	NUMBERS	JOURNALS	BUFFERS	BUFFER HEADS	COUPLING	BRAKE
2554	450	B731590 TO B732039	10"x 5" PLAIN	SPINDLE 1'-6" PROJECTION	1'-4" DIA.	3 LINK	HAND LEVER
2740	350	B732040 TO B732389	10"x 5" PLAIN	SPINDLE 1'-6" PROJECTION	1'-4" DIA.	INSTANTER	A.V.B. CLASP & HAND LEVER
2867	650	B732390 TO B733039	10"x 5" PLAIN	SPINDLE 1'-6" PROJECTION	1'-4" DIA.	INSTANTER	A.V.B. CLASP & HAND LEVER
3226	180	B733040 TO B733219	4⅞" DIA. ROLLER	SELF CONTAINED 1'-6" PROJECTION	1'-6" DIA.	INSTANTER	A.V.B. CLASP & HAND LEVER
PART 3288	150	B730500 TO B730649	9"x 5" PLAIN	SELF CONTAINED 1'-6" PROJECTION	1'-6" DIA.	INSTANTER	A.V.B. CLASP & HAND LEVER
PART 3288	270	B730650 TO B730919	9"x 5" PLAIN	HYDRAULIC 1'-8½" PROJECTION	1'-6" DIA.	INSTANTER	A.V.B. CLASP & HAND LEVER
3332	300	B730920 TO B730999 B733240 TO B733459	9"x 5" PLAIN	HYDRAULIC 1'-8½" PROJECTION	1'-6" DIA.	INSTANTER	A.V.B. CLASP & HAND LEVER

Over 2,000 Tube wagons to Diagram 1/448 were built at Darlington and Derby between 1954 and 1961. Different sizes of buffer head and buffer length existed and whilst the first batch were only hand braked, later deliveries were vacuum fitted. Unlike Plates, where minor variations called for a separate diagram, one diagram was deemed adequate to cover a multitude of differences.

12 TON STRIP COIL WAGON

CONVERTED FROM 13 TON PALLET BRICK WAGON TO B.R. DIAGRAM No. 23

Scale: 4mm = 1 ft

BR1/452

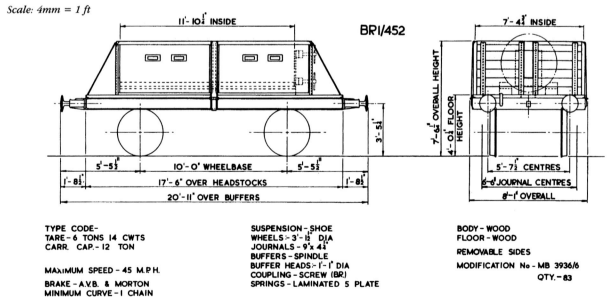

TYPE CODE –
TARE – 6 TONS 14 CWTS
CARR. CAP. – 12 TON

MAXIMUM SPEED – 45 M.P.H.
BRAKE – A.V.B. & MORTON
MINIMUM CURVE – 1 CHAIN

SUSPENSION – SHOE
WHEELS – 3'-1½" DIA
JOURNALS – 9" x 4¼"
BUFFERS – SPINDLE
BUFFER HEADS – 1'-1" DIA
COUPLING – SCREW (BR)
SPRINGS – LAMINATED 5 PLATE

BODY – WOOD
FLOOR – WOOD
REMOVABLE SIDES
MODIFICATION No – MB 3936/6
QTY. – 83

Seven separate diagrams were issued for the 1,420 Palbricks built between 1954 and 1961, though they were coded as Palbrick A, B or C. Within a few years the decline in brick traffic had resulted in many being surplus to requirements, and some enjoyed a fresh lease of life as steel coil carriers. Of the 380 Diagram 1/023 Palbrick A's, (B461617–461996), which were built at Ashford,

83 were later fitted with wooden cradles to enable them to carry strip coils, a traffic which was increasing rapidly in the late 'fifties and 'sixties, and was to see many types of wagon converted to meet the burgeoning need for Coils.
A new Diagram, 1/448, was issued for the new Coil Q, as these wagons were known.

42 TON BOGIE BOLSTER

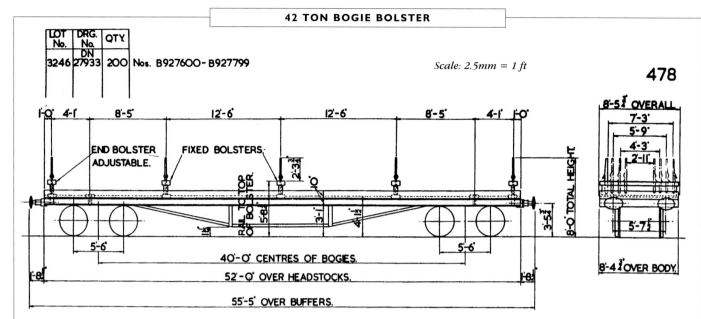

LOT No.	DRG. No. DN	QTY.
3246	27933	200

Nos. B927600 - B927799

Scale: 2.5mm = 1 ft

478

TEL. CODE :- BOBOL D.
TARE :- 22 TONS 1 CWT.
CARRYING CAPACITY :- 42 TONS.
BINDING CHAINS INCLUDING SCREW.

FLOOR :- WOOD.
MINIMUM CURVE :- 70 FT.
HAND SCREW & A.V. BRAKE WITH S.A.B. REGULATOR.

WHEELS :- 3'-1½' DIA.
JOURNALS :- 10'x 5'
BUFFERS :- 24'x14'
COUPLINGS :- INSTANTER.

As with the majority of BR stock, the 42 ton Bogie Bolster D was derived from LMS/LNER practice. Over 3,000 entered service between 1949 and 1962 when construction ended. They were built to five separate diagrams, 150 being to the LMS inspired Diagram 1/470, and some 2,000 to its LNER counterpart, 1/472. Until 1958, all were unfitted, but the last 200 examples to

Diagram 1/472 received vacuum brakes. A new Diagram, 1/478, was issued in 1959, when 200 more Bolster Ds emerged from Lancing with shortened stanchions and vacuum brakes.
The design perpetuated the deep LNER-style composite bolster. The diagram reveals that only the end bolsters were movable and that the stanchions could be placed in any of four positions.

42 TON TRESTLE WAGON

Scale: 2mm = 1 ft

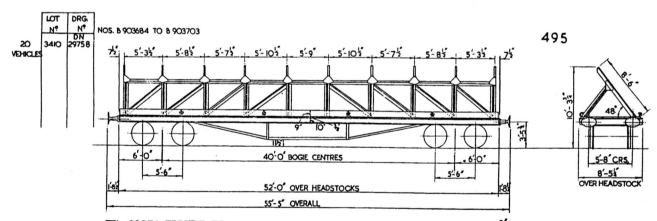

LOT Nº	DRG. Nº DN
3410	29758

20 VEHICLES

NOS. B 903684 TO B 903703

495

TEL. CODE :- TRESTLE E.D.
TARE :- 20 TONS 15 CWTS.
COUPLINGS :- INSTANTER
MINIMUM CURVE :- 70'-0'
BRAKE. HAND LEVER & A.V.B.
WITH S.A.B. REGULATOR

JOURNALS :- 4⅛' DIA. ROLLER BEARINGS.
DIA. OF WHEELS :- 3'-1½'
BUFFERS :- SELF CONTAINED
BUFFER HEADS :- 24' X 14' OVAL.
BOGIE :- CAST STEEL.

The Diagram 2/495 Trestle wagon was one of a small group of bogie bolster wagon chassis completed with sloping cradles for the conveyance of steel plate which

would be too wide if laid horizontally on the wagon floor. Twenty (B903684-903703) roller-bearing Trestle EDs were built to Diagram 2/495.

Scale: 4mm = 1 ft

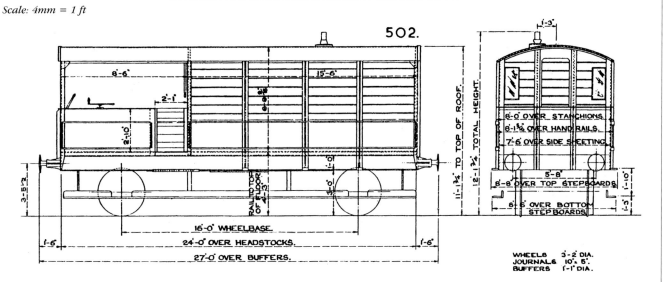

502.

TEL. CODE:- TOAD.
TARE. 20 TONS.
MIN. CURVE 1½ CHAIN.
SCREW HAND BRAKE.

For the modeller who wants something slightly different, here is the answer, a British Railways Toad. Given the GWR belief in the superiority of all things Great Western, it is hardly surprising that Swindon managed to turn out a final batch of 74 traditional Toads in 1949, disguising them as BR diagram 1/502. The vans, B950542-950615, carried a NOT IN COMMON USE branding and were normally kept on ex-GWR metals, but are known to have made occasional forays into alien territory.

---

20 TON GOODS BRAKE VAN

Scale: 4mm = 1 ft

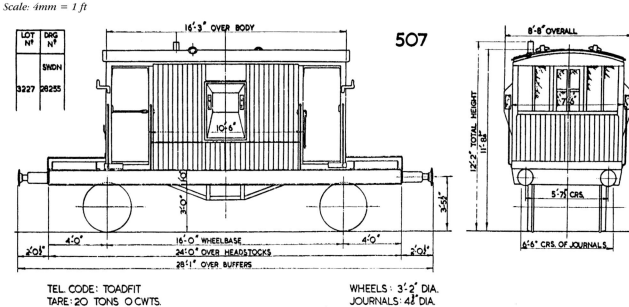

507

TEL. CODE: TOADFIT
TARE: 20 TONS O CWTS.
MIN CURVE: 1½ CHAINS
BRAKE: SCREW HAND BRAKE
BUFFERS: 2'-0½" PROJECTION.
BUFFER HEADS: 1'-1" DIA.
FITTED A.V.P & GAUGE.

WHEELS: 3'-2" DIA.
JOURNALS: 4⅛" DIA.
ROLLER BEARINGS.
COUPLINGS: R.I.V. SCREW.

The BR standard brake van closely followed LNER practice, and appeared to four separate diagrams. Diagrams 1/500 and 1/504 copied LNER practice, even to the short footsteps, but Diagram 1/506 introduced full length footsteps.

In Diagram 1/507, roller bearings and hydraulic buffers became standard. 727 vans (B954521-955247) were built to this design for BR, whilst a further six, (L580-585), were completed for London Transport.

14 TON BALLAST SIDE TIPPING WAGON

Scale: 4mm = 1 ft

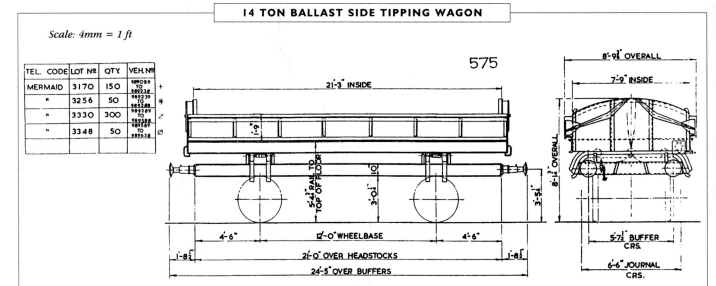

575

TEL. CODE	LOT Nº	QTY	VEH.Nº	
MERMAID	3170	150	989089 TO 989238	+
"	3256	50	989239 TO 989288	*
"	3330	300	989289 TO 989588	∠
"	3348	50	989589 TO 989638	∅

TARE :- 10 TONS 19 CWTS.
CUBIC CAPACITY :- 15 CU. YDS.
MIN. CURVE :- 1¼ CHAINS.
BRAKE :- A.V.B. & HAND LEVER.
DOORS :- 2 FULL BODY LENGTH SIDE DOORS.

BODY :- STEEL.
FLOOR :- STEEL.
DIA. WHEELS :- 3'-1½".
JOURNALS :- 9"x4½".

BUFFING & DRAWGEAR.
+ I.R. SELF CONTAINED (1'-4"DIA. HEAD) SCREW COUPLING.
* HYDRAULIC (1'-1"DIA. HEAD) SCREW COUPLING.
∅ HYDRAULIC (1'-1"DIA. HEAD) INSTANTER.

After an introduction to the esoteric delights of inches, feet, yards and miles, my wife accepted that the English might call wagons SEA HARE or BARBEL, but was somewhat surprised when I told her that some wagons had once been called 'Rusalka'. To be truthful, they did not actually carry the Russian version of the name, but its English equivalent, which is Mermaid!

Apart from Great Western Mermaids, which were entirely plausible creatures, given the legends enveloping the Cornish Riviera, two varieties of nationalised Mermaid existed, the Diagram 1/473 unfitted Mermaid, and the Diagram 1/475 vacuum braked Mermaid. Four lots of fitted Mermaid were built by Metro-Cammell, with three different types of buffing and drawgear!

24 TON HOPPER BALLAST WAGON

Scale: 4mm = 1 ft

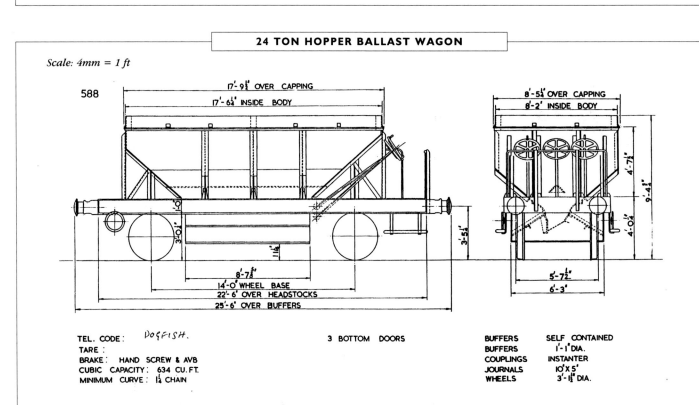

588

TEL. CODE :	DOGFISH.			
TARE :		3 BOTTOM DOORS	BUFFERS	SELF CONTAINED
BRAKE : HAND SCREW & AVB			BUFFERS	1'-1"DIA.
CUBIC CAPACITY : 634 CU. FT.			COUPLINGS	INSTANTER
MINIMUM CURVE : 1¼ CHAIN			JOURNALS	10"X 5"
			WHEELS	3'-1¼"DIA.

The Diagram 1/588 Dogfish offers a chance to customise standard wagons, as it is a rebuild of the usual Diagram 1/587 Dogfish with a shallow upward extension to increase the carrying capacity to 634 cubic feet for the conveyance of

slag ballast. One of these wagons, No DB992838, rebuilt to Diagram 1/588, is illustrated on page 72. One such wagon included amongst a rake of unmodified wagons would add interest to a layout.

Scale: 4mm = 1 ft

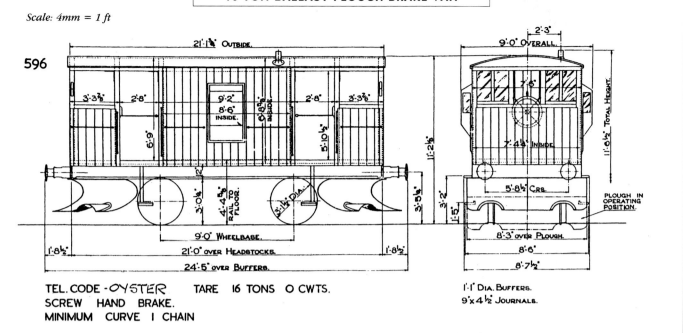

596

TEL. CODE - OYSTER   TARE 16 TONS 0 CWTS.
SCREW HAND BRAKE.
MINIMUM CURVE 1 CHAIN

1'-1" DIA. BUFFERS.
9" x 4½" JOURNALS.

Every schoolboy knows that a tadpole becomes a frog, but how many people are aware that a Shark could turn into an Oyster? In the enchanting world of departmental vehicles, such transformations can take place at the stroke of a pen. Nine of these remarkable creatures existed, DB993700-993708. They were a copy of the LMS 16 ton ballast plough van, the design being beefed up to a 20 ton vehicle for later deliveries.

After many years as Sharks, they underwent a mutation hitherto unknown to the marine biologist, but whether the new name was carried on the wagons, rather than just in the diagram books is uncertain, as Oyster was one of the names allocated to the Diagram 1/641 40 ton rail wagon. Equally confusingly, this was also known in the diagram books as Gane A.

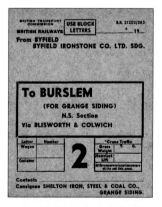

A further selection of wagon labels. Pre-nationalisation labels remained in use for some years, eg. LMS Live Stock. Empty Minfit was a late period development reflecting the growing number of vacuum-fitted 16 tonners.
The remaining examples cover special traffic and the ubiquitous 'anything anywhere' label.

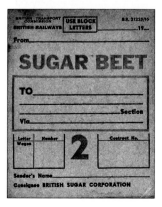

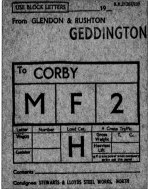

GLOSSARY

Bell Code — Block Telegraph code used in semaphore signalling installations to offer trains from box to box.

Bolster — Wagon with transverse baulks to support long loads such as rails or timber.
Term also used for the timber baulk on which the load rests.

Borail — Special bogie wagon for rail or other long loads.

Catfish — Departmental stock code for 19 ton ballast hopper.

Chauldron — Name given to early coal wagons with slope sides and bottom doors. They could carry one Chauldron or Chaldron of coal, a measure which differed from area to area (eg. London or Newcastle Chaldron) but was about 53 cwt (2.65 tons).

Clasp Brakes — 8 shoe brake system with shoes clasping opposite sides of a wheel's tyre.

Commuted Charge — See page 55 for details.

Conflat — A 4-wheel flat wagon used to convey containers.

Covhop — A covered hopper wagon.

Curb Rail — Also known as a side rail, a longitudinal member above the solebar on which body planks rest.

Diagram — The page number in official diagram books showing outline drawings of vehicle types.

Dog Clutch — A reversing mechanism whereby brakes can be applied from either side of a wagon.

Dogfish — Departmental stock code for a 24 ton ballast hopper.

Dumb Buffers — Solid wooden buffers fitted to early wagons.

Fitted Stock — Wagons with continuous brakes; until 1968 this meant vacuum brakes.

Flatrol — Dropped centre bogie wagon with load resting on floor of wagon.

Flitch Plates — Iron or steel plate on face of solebar.

Girdwag — BR name for wagons to convey very heavy loads such as naval guns or girders, see also Pollen.

GLW — Gross Laden Weight, the maximum permitted weight of wagon plus load.

Grampus — Departmental stock code for BR dropside/movable end ballast wagons.

Gulf Red — Short lived departmental red colour scheme for Engineers' stock.

Haddock — Departmental stock code for a sleeper wagon with drop sides.

Headstocks — Sometimes called the buffer beam to which buffers and couplings are attached.

Herring — Departmental stock code for a 20 ton ballast hopper.

High — High sided 4, 5 or 5½ -plank general merchandise open wagon.

High Steel — LNER steel bodied version of wooden bodied High, also built by BR.

Hopper — Self-discharging wagon with sloped sides leading to a bottom door.

Instanter Couplings — A 3-link coupling with a central link capable of being put in long or short positions to reduce snatching/jerking of wagons during braking or acceleration.

Insulfish — LNER/BR insulated fish van.

Load — The maximum permitted load a wagon was allowed to carry, given in tons.

Loads Book — BR instructions to staff on freight train operating.

Lowfit — Low sided general merchandise wagon with vacuum brakes, often used for end loading.

Lowmac — 4-wheel wagon with a lowered centre section often used to transport tall road vehicles.

Machinery Truck — 4-wheel wagon with end doors used to move farm implements, cars and other wheeled items.

Match Wagon — Used when long loads (eg timber) overhangs the end of a wagon.

Medfit — Medium sided general merchandise open wagon with vacuum brakes.

Medium — Medium sided open with hand brakes.

Mermaid — Departmental stock code for 14 ton side tipping ballast hopper.

Mica — GWR telegraph code for an insulated meat van.

Mink — GWR telegraph reporting code for a 4-wheel van.

Minfit — Fitted mineral wagon.

Monobloc — Integral construction tank wagon with dropped frame members.

Monster — GWR telegraph code for bogie van.

Morton Brakes — A system whereby brakes can be applied from either side of a wagon (see also Dog clutch).

MOWT — Ministry of War Transport.

Open — General name for open merchandise wagon.

Palbrick — Pallet wagon for brick traffic also known as a Palwag.

Palvan — 4-wheel BR van with an extra wide door at one end for pallet traffic.

Palwag — See Palbrick above.

Pipe — Long wheelbase 4-wheel open wagon for pipes or similar loads.

Plate — Long wheelbase 4-wheel open wagon for steel plate or similar loads.

Pollen — GWR telegraph code for wagons used to convey heavy loads such as large naval guns.

PO/POW — Private owner wagon, usually coal or tank wagons.

Presflo — Covered hopper cement wagon with air pressure assisted discharge.

RCH — Railway Clearing House; set up by the railway companies to handle common interests.

Shark — Departmental stock code for a ballast plough / brake van.

Shoc — Railway reporting code for Shock wagon, also Shocbar, Shocvan etc.

Shock — Van or open with longitudinal springs between chassis and body to absorb shunting shocks.

Sole — Departmental stock code for 3-plank dropside/drop end ballast wagon.

Solebars — Outside longitudinal frame members.

Specially Constructed Vehicles — Wagons designed for unusual loads.

Stanchion — Vertical post as on a bolster wagon.

Strapping — An iron or steel strip to which body planks are bolted.

Sturgeon — Departmental stock code for a 50 ton bogie rail and sleeper wagon.

Tare Weight — The weight of a wagon when empty.

Tippler — A wagon without doors which was emptied by being inverted in a rotary tippler.

Toad — GWR telegraph code for a goods brake van.

TOPS — Total Operations Processing System; a computerised stock location and identification system introduced by BR in 1972.

Transformer MB/MC — A 120/135 ton trolley wagon to move large transformers.

Trout — Departmental stock code for a 25 ton ballast hopper.

Unfitted Stock — Wagons with hand brakes only.

V Hanger — Central V shaped hanger which supports a brake shaft.

Vanwide — 4-wheel BR development of the Palvan with a wide central opening door.

Warflat — Bogie flat wagon used to move tanks and other military vehicles.

Warwell — Bogie wagon developed during the Second World War to move tanks or other military vehicles.

Weltrol — A large bogie wagon with its centre lower than the end, the load rests on the side girders of the wagon.